VEHICLE INSPEC

HOW SAFE IS YOUR CAR?

An MOT handbook on mechanical safety

also covers
light vans and pick-ups

Developed and prepared by COMIND
Written by The Vehicle Inspectorate
Edited by Nick Lynch
Illustrated by Stephen Oakman and Vickey Squires

First Published 1990
Second edition 1994

ISBN 0 11 5511652

Foreword

Road safety matters. It is important to pedestrians and other road users as well as to individual owners and drivers. Motorists must share in this responsibility, which means driving carefully in a roadworthy vehicle and observing the requirements of the law.

For your vehcile to be safe and roadworthy, it must be regularly checked and maintained. Cars over three years old have to undergo the 'MOT' test every year, which is aimed at ensuring compliance with basic safety, legal and environmental requirements.

The practical, regular checks set out in this book are based on the 'MOT' test requirements. They are a handy guide to help you look after your car. Safe driving!

Robert Key MP
Minister for Roads and Traffic

Important notes

1. Regulations

This book is based on the regulations at the time of going to press. New regulations will come into force from time to time, or existing regulations may be amended.

Information on the up-to-date regulations is available from MOT Testing Stations and from the Vehicle Inspectorate.

2. Tolerances

Because it is not possible to specify exact tolerances for wear, play, corrosion and so on, words such as 'excessive', 'serious' and 'too much' are sometimes used. If you feel unable to judge any item safely, seek expert advice.

INTRODUCTION

Why motorists should use this book

This book is produced by the Vehicle Inspectorate. Its purpose is to give motorists easy-to-follow advice on how to keep their vehicles safe, in good condition and within the legal requirements of the 'MOT' Test.

It is not intended to interpret the law governing the 'MOT' Test: that's a matter for the courts to determine.

This book does not replace the existing official MOT Testers Manual — Vehicle Testing used in the Test and also available from HMSO. Some testable items, such as brakes and headlamp-beam aim, can be difficult for you to check effectively without proper equipment.

This means that your vehicle can still fail, although you might have followed the procedures in this book to the letter. Remember that the 'MOT' tester is specially trained and experienced: pass or fail is at his discretion.

If you feel you have a genuine grievance over a failure, you can appeal to the Vehicle Inspectorate's local Enforcement Area Office (see page 110).

While this book has been put together as a manual for those who want to check and/or prepare their vehicles for the 'MOT' Test, it makes good sense to follow regularly the routines set out. Even if your vehcile is not due for an 'MOT' Test, or perhaps not even within the scope of the scheme, it will pay you to carry out all these checks as part of your routine maintenance.

Put right any defects straight away: a minor defect can become dangerous if you neglect it.

If you have any doubts, don't hesitate to seek expert advice.

CONTENTS

ROUTINE

Follow these steps to check your car

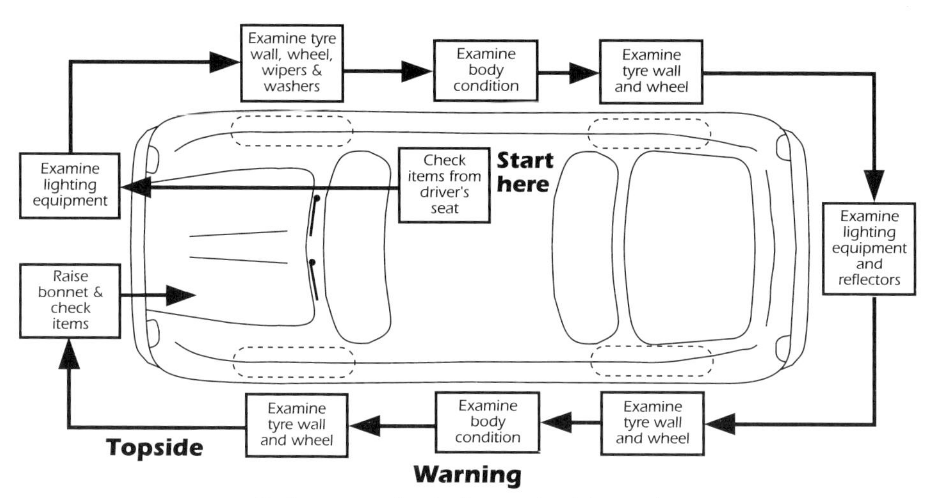

Warning

Before you check the underside, make sure your car is safely jacked up and secured

With the front wheels in the straight ahead position, check all relevant underside items starting at offside front

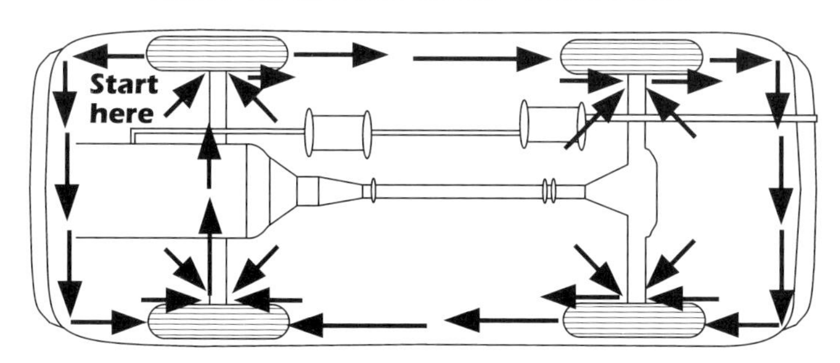

Underside

HEADLAMPS - 1

Are your headlamps working correctly?

WHAT THE MOT REQUIRES	HOW TO CHECK
The headlamps **Your car must have:** 1. Two **dipped-beam** headlamps 2. Two **main-beam** headlamps Cars used only in daylight do not require lamps. You must either ▪ remove **all** lamps, or ▪ disconnect them permanently, or ▪ paint over them, or ▪ mask them **Cars requiring no headlamps** Cars first used before 1 January 1931 do not require headlamps **Cars requiring only one headlamp** All three-wheeled vehicles first used before 1 January 1972 Three-wheeled first used on or after 1 January 1972 ▪ weighing less than 400kg, and ▪ less than 1.3m wide **Front fog lamps** Front fog lamps or auxiliary lamps are not included in the Test	**With ignition on, switch on lights and make sure that:** a. the lamps show a steady white or yellow light visible from a reasonable distance b. the light of any lamp is not affected when you switch on another lamp c. no lamp flickers when tapped d. each lamp is secure e. no lens is damaged or missing (for cracks see opposite page) f. all switches work correctly g. both matching lamps (eg both headlamps) show the same colour h. the aim of the headlamps complies with the relevant diagram on pages 6 and 7 i. matching pairs of lamps are in the correct position

Are your headlamps safe?

MAIN REASONS FOR FAILURE	REMARKS	ACTION	✓
1. Two symmetrically placed headlamps NOT ▪ showing a steady WHITE or YELLOW light ▪ clean and in good working order	⚠⚠	Check Repair wiring/ Replace Recharge battery/ check positioning	
2. A lamp ▪ does not light up as soon as it is switched on ▪ is affected by the working of another lamp ▪ flickers when you tap it	⚠⚠⚠	Check wiring and connections Repair/Replace	
3. A lens ▪ missing ▪ damaged (see below)	⚠	Repair/Replace	
4. A loose lamp	⚠⚠	Tighten	
5. A faulty switch	⚠⚠	Repair/Replace	

WARNING

Extremely dangerous. DO NOT drive your car in this condition.
You will be breaking the law and risking your life and the lives of others.

Very dangerous and may be illegal. Put right immediately.

Could also be dangerous and illegal.

Note: A 'matched pair of lamps' means

- a pair of lamps the same height above ground
- one lamp on each side of car equal distance from edge of car
- both showing the same colour light

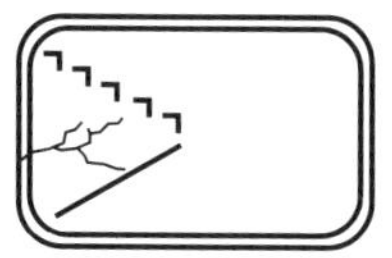

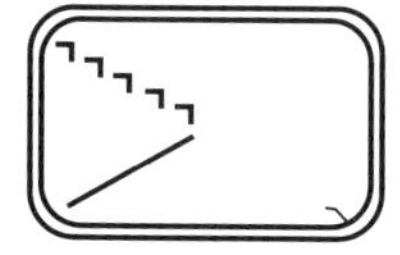

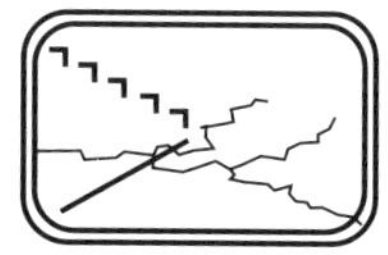

Cracks in headlamp lenses (✓ acceptable ✗ not acceptable)

HEADLAMP AIM - 1

HOW TO TEST YOUR HEADLAMPS

You will need:

- A piece of chalk or other suitable marker
- A 2 metre tape measure
- A long straight edge
- A reasonably level piece of ground and a wall
- An assistant

The test:

1. Check that your tyres are correctly inflated.
2. Position your car squarely up to and as close to the wall as possible.
3. Find the centre of each headlamp lens and mark corresponding crosses on the wall.
4. Move the car back squarely so that the headlamp lenses are 2 metres from the wall.
5. Draw a straight line between the two crosses and mark a line vertically downwards from each cross.
6. Switch on the headlamps 'main' and 'dipped' beams in turn. One of the beams will match one of the three images shown in the diagram on pages 6 and 7. The lens markings shown above each diagram may help identify the type of headlamp fitted.
7. Check that the boundaries of the beam image are not outside the limits given on the appropriate headlamp diagram for each lamp in turn. Ask your assistant to sit in the driver's seat and guide you on the location of the beam boundaries.
8. Your Owner's Handbook should tell you how to adjust the headlamp aim if it is wrong.

Note: This check will give a good indication on the alignment of your vehicles headlamps, but remember that MOT testing stations have properly calibrated equipment and level floors to check headlamp aim accurately.

HEADLAMP AIM - 2

distance between headlamp centres

height of headlamps above floor

horizontal line

floor

Drawing on wall

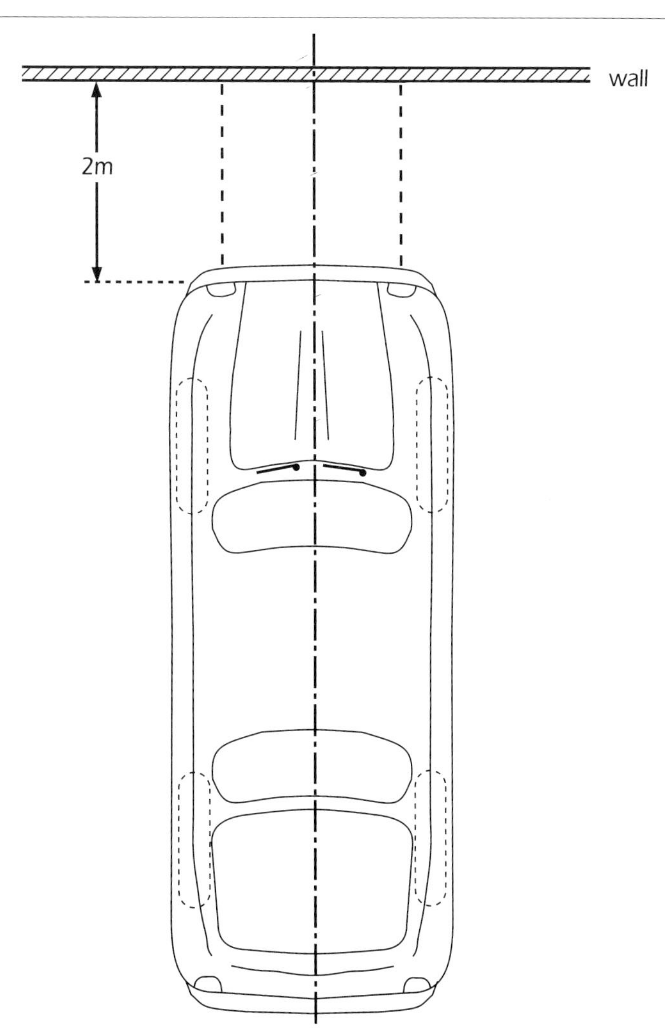

Plan of car

HEADLAMP AIM - 3

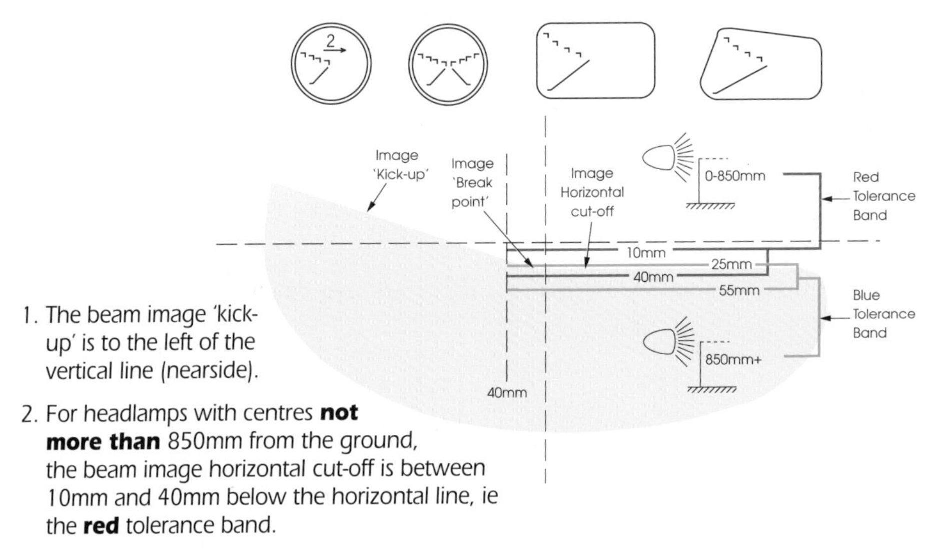

1. The beam image 'kick-up' is to the left of the vertical line (nearside).
2. For headlamps with centres **not more than** 850mm from the ground, the beam image horizontal cut-off is between 10mm and 40mm below the horizontal line, ie the **red** tolerance band.
3. For headlamps with centres **more than** 850mm from the ground, the beam image horizontal cut-off is between 25mm and 55mm below the horizontal line, ie the **blue** tolerance band.
4. The beam image 'break point' is to the left of the vertical line, but not by more than 40mm.

Diagram 1.
European type headlamps checked on dipped beam

HEADLAMP AIM - 4

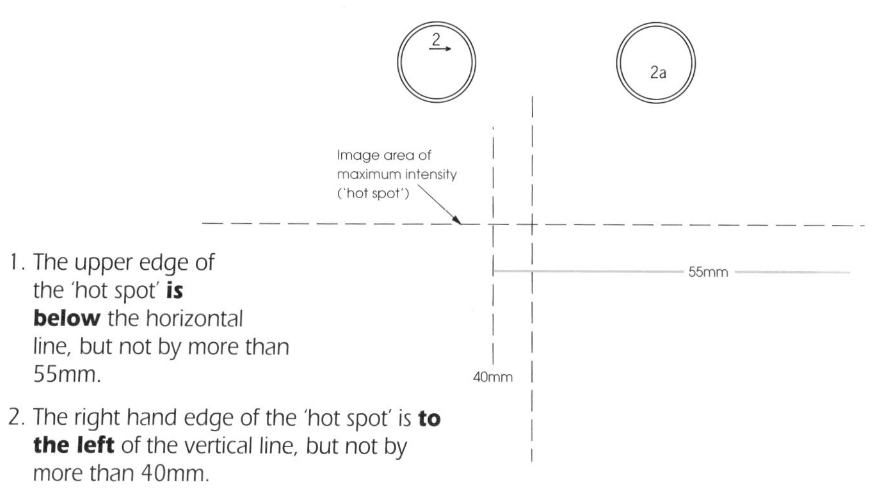

1. The upper edge of the 'hot spot' **is below** the horizontal line, but not by more than 55mm.
2. The right hand edge of the 'hot spot' is **to the left** of the vertical line, but not by more than 40mm.

Diagram 2.
British American type headlamps checked on dipped beam

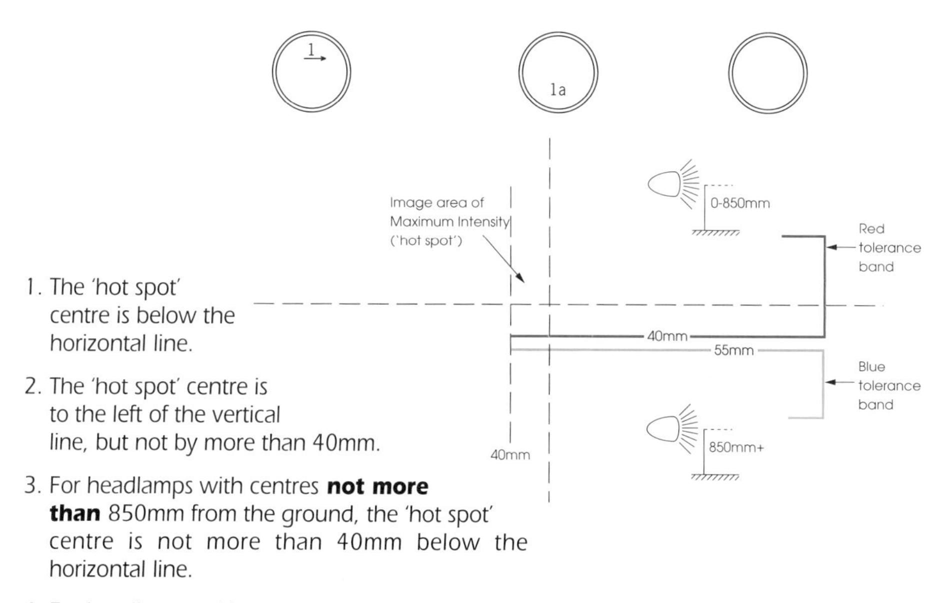

1. The 'hot spot' centre is below the horizontal line.
2. The 'hot spot' centre is to the left of the vertical line, but not by more than 40mm.
3. For headlamps with centres **not more than** 850mm from the ground, the 'hot spot' centre is not more than 40mm below the horizontal line.
4. For headlamps with centres **more than** 850mm from the ground, the 'hot spot' centre is not more than 55mm below the horizontal line.

Diagram 3.
British American type headlamps checked on main (driving) beam

FRONT & REAR POSITION LAMPS - 1

Are your front lights safe?

WHAT THE MOT REQUIRES	HOW TO CHECK
Your car must have: 1. **Two** front position lamps (side lamps) which give out a steady WHITE light (or a YELLOW light if incorporated into a headlamp giving out yellow light) 2. **Two** rear position lamps which show a steady RED light These lamps must ▪ be clean and in good working order ▪ show a steady light visible from a reasonable distance ▪ be equal distance from centre of car ▪ be about equal height above the ground **Cars which do not require lamps** Cars used only in daylight do not require lamps. You must either remove all lamps or disconnect or mask them	**Switch on lights and make sure that:** a. front position lamps show a WHITE light (or a YELLOW light: see What the MOT Requires — 1) b. rear position lamps show a steady RED light c. no lamp is affected by the operation of any other lamp d. no lamp flickers when tapped e. each lamp is secure f. no lens is damaged or missing (for cracks, see page 3) g. all switches work correctly h. the light is visible from a reasonable distance i. matching pairs of lamps are in the correct position

FRONT & REAR POSITION LAMPS - 2

Are your rear lights safe?

MAIN REASONS FOR FAILURE	REMARKS	ACTION	✓
1. Two symmetrically placed ▪ front position lamps which do not show a steady WHITE light (or a YELLOW light) ▪ rear position lamps which do not show a steady red light The light is not visible from a reasonable distance	⚠ ⚠	Check bulb Check wiring Repair/Replace	
2. A front or rear position lamp ▪ does not light up as soon as it is switched on ▪ is affected by the working of another lamp ▪ flickers when you tap it ▪ loose	⚠ ⚠ ⚠	Check wiring and connections Repair/Replace/ Tighten	
3. A lens ▪ missing ▪ damaged (see page 3)	⚠	Replace	
5. A switch ▪ faulty ▪ missing	⚠ ⚠	Repair/Replace	

WARNING

Extremely dangerous. DO NOT drive your car in this condition.
You will be breaking the law and risking your life and the lives of others.

Very dangerous and may be illegal. Put right immediately.

Could also be dangerous and illegal.

STOP LAMPS - 1

Are both your stop lamps working?

WHAT THE MOT REQUIRES	HOW TO CHECK
Your car is required to have two stop lamps ▪ visible from the rear at a reasonable distance ▪ which give out a steady RED light when the footbrake pedal is pressed Cars not required to have stop lamps are those ▪ first used before 1 January 1936 ▪ with a maximum speed not exceeding 15mph (by Law or power) ▪ without electrically operated lighting equipment Cars required to have only **one** stop lamp are those first used on or after 1 January 1936 and before 1 January 1971	Ask a friend to sit in the driver's seat and press and release the footbrake alternately as you instruct. 1. Check that both stop lamps show RED light visible from a reasonable distance 2. Check that the red light is steady and stays on while footbrake is applied 3. Check that stop lamps do not show a red light when the footbrake is NOT APPLIED 4. Check lens for ▪ serious cracks ▪ serious chipping ▪ missing fragments of glass

STOP LAMPS - 2

Are both your stop lamps in good condition?

MAIN REASONS FOR FAILURE	REMARKS	ACTION	✓
1. Stop lamps do not show a RED light visible from a reasonable distance	⚠⚠⚠	Check bulb Check wiring Check switch Repair/Replace	
2. Stop lamps do not stay steadily lit while brakes are applied	⚠⚠	Check wiring Repair/Replace	
3. Stop lamps remain on while brakes are not applied	⚠	Check wiring Check switch Repair/Replace	
4. A stop lamp ▪ missing ▪ insecure	⚠⚠	Repair/Replace	
5. A lens ▪ badly repaired ▪ missing ▪ damaged so that it does not do its job	⚠	Repair/Replace	
6. A lens is ▪ seriously cracked ▪ has missing fragments of glass	⚠	Replace	

WARNING

Extremely dangerous. DO NOT drive your car in this condition.
You will be breaking the law and risking your life and the lives of others.

Very dangerous and may be illegal. Put right immediately.

Could also be dangerous and illegal.

INDICATORS & HAZARD WARNING LAMPS - 1

Are your indicators working correctly?

WHAT THE MOT REQUIRES	HOW TO CHECK
The indicators Your car must have two direction indicators which show an amber light to the front, and an amber light to the rear They must ▪ flash between 60-120 times per minute (with engine running, if necessary) ▪ be clean and in good working order If your car is not fitted with any front and rear position lamps, it is not required to have indicators and hazard warning lamps. If your car has indicators and hazard warning lights fitted they will be tested. Cars first used ▪ on or after 1 April 1986, must have a side indicator repeater on each side **Note:** The side repeater might be part of the front direction indicator if it includes a wrap-around lens. ▪ before 1 April 1986, are not required to have hazard warning lights ▪ before 1 January 1936, are not required to have direction indicators ▪ before 1 September 1965, can have white indicators on the front and red indicators on the rear. For cars fitted with semaphore arms, they should ▪ move freely and not stick ▪ be illuminated amber on both sides ▪ be seen by driver from a normal driving position. If not, an audible or visible telltale must be fitted and be working correctly.	**Indicators** 1. Operate the left and right direction indicators in turn and check that all required lamps are a. present b. complete c. secure 2. Time the flashing rate (you might have to have the engine running for this). The lights must flash 60-120 times per minute. 3. Check that each indicator shows an amber light and that the 'tell-tale' is working correctly. 4. Make sure that the working of the indicator lamps is separate from, and does not interfere with, the working of any other lamps. **Hazard Warning Lights** 1. Check that the hazard warning device operates with the ignition switched both off and on. While it operates, check that all the direction indicators flash simultaneously and the 'tell-tale' is working correctly.

INDICATORS & HAZARD WARNING LAMPS - 2

Are your hazard warning lamps working?

MAIN REASONS FOR FAILURE	REMARKS	ACTION	✓
1. An indicator missing or not visible	⚠ ⚠	Repair/Replace	
2. An indicator lamp ▪ not the correct colour ▪ not visible from a reasonable distance	⚠ ⚠	Repair/Replace	
3. An indicator ▪ not working ▪ insecure ▪ not flashing at the correct rate	⚠ ⚠	Check wiring Repair/Replace	
4. An indicator lamp ▪ does not light up as soon as it is switched on ▪ connected to the working of another lamp ▪ is affected by the working of another lamp	⚠ ⚠	Check switch Check wiring and connections Repair/Replace	
5. A lens ▪ badly repaired ▪ missing ▪ damaged so that it does not do its job	⚠	Repair/Replace	
6. An indicator lamp ▪ loose ▪ flickers when you tap it	⚠ ⚠	Tighten/Repair	
7. A switch ▪ missing ▪ loose	⚠ ⚠	Replace	
8. A tell tale ▪ not working ▪ not visible/audible	⚠	Repair/Replace	
9. A hazard warning device ▪ does not cause all the direction indicators to flash in phase ▪ 'tell-tale' not working correctly	⚠	Repair/Replace	

WARNING

Very dangerous and may be illegal. Put right immediately.

Could also be dangerous and illegal.

REAR FOG LAMPS - 1

Is your rear fog lamp working?

WHAT THE MOT REQUIRES	HOW TO CHECK
If your car was first used on or after 1 April 1980, it must have one rear fog lamp positioned either ▪ in the centre of the vehicle ▪ on the offside Extra rear fog lamps will not be tested. Only the lamp in the required position must be working. If there is a second lamp, for example, on the nearside, you should be sure that it is working also to give you extra protection in heavy fog.	With ▪ the engine switched on, and ▪ dipped headlights Check that ▪ the rear fog lamp fitted to the CENTRE or to the OFFSIDE gives off an intense RED light ▪ the driver's telltale light works correctly

REAR FOG LAMPS - 2

Is your rear fog lamp in good condition?

MAIN REASONS FOR FAILURE	REMARKS	ACTION	✓
1. A rear fog lamp ▪ missing ▪ does not show a steady RED light ▪ is not visible from a reasonable distance		Fit one Check wiring	
2. A rear fog lamp ▪ loose ▪ incomplete ▪ flickers when you tap it ▪ is effected by the operation of another lamp		Tighten/Repair	
3. A rear fog lamp tell tale does not work		Check wiring/ tell tale lamp	

WARNING

Extremely dangerous. DO NOT drive your car in this condition.
You will be breaking the law and risking your life and the lives of others.

Very dangerous and may be illegal. Put right immediately.

NUMBER PLATE LAMPS - 1

Is your registration number clear at night?

WHAT THE MOT REQUIRES	HOW TO CHECK
Your car must have at least one number plate lamp on the rear number plate which lights up when the front/rear position lamps are switched on. If your car number plate has more than one light, all lights must be working correctly.	With the front and rear position lamps working, check that ALL lamps on the rear number plate are ▪ present ▪ working correctly ▪ secure

NUMBER PLATE LAMPS - 2

MAIN REASONS FOR FAILURE	REMARKS	ACTION	✓
1. A number plate lamp ▪ not fitted ▪ not working correctly ▪ flickers when you tap it	⚠ ⚠	Fit one Repair/Replace	

WARNING

Could also be dangerous and illegal.

REAR REFLECTORS - 1

Are your reflectors secure?

WHAT THE MOT REQUIRES	HOW TO CHECK
The rear reflectors **Your car must have:** 1. Two rear retro reflectors that are ▪ unobscured ▪ fitted squarely to face the rear ▪ equal distance from edge of car, one on each side at rear ▪ clean Cars used only during the day that ▪ have no lamps ▪ have disconnected or masked lamps do not need reflectors Extra reflectors are not included in the Test **NOTE: Reflective tape is NOT acceptable in place of a reflector**	**Check that your car has clean RED rear reflectors** Examine their condition, security and position

REAR REFLECTORS - 2

Are your reflectors in good condition?

MAIN REASONS FOR FAILURE	REMARKS	ACTION	✓
1. A rear reflector ▪ missing ▪ obscured ▪ incorrectly positioned ▪ not red	⚠⚠⚠	Fit one/ Repair/Replace	
2. A reflector ▪ damaged ▪ not doing its job	⚠⚠	Repair/Replace	
3. A loose reflector	⚠	Tighten	

WARNING

Extremely dangerous. DO NOT drive your car in this condition.
You will be breaking the law and risking your life and the lives of others.

Very dangerous and may be illegal. Put right immediately.

Could also be dangerous and illegal.

STEERING - 1

Checks from driver's seat

Is your car's steering safe?

WHAT THE MOT REQUIRES	HOW TO CHECK
The steering controls The condition of the STEERING must not cause danger to any person in the car or on the road The STEERING must be in good and efficient working order and be properly adjusted	Release steering lock, if car has one a. Check that any mechanism for adjusting steering column is fully locked b. Turn steering wheel from side to side while checking for excessive movement due to wear c. Pull and push steering wheel while checking for movement in centre of column d. Check that steering column is securely mounted to body of car e. Check steering wheel hub, spokes and rim for fractures or loose spokes

Checks from driver's seat

Is your car's steering wheel/column in good condition?

MAIN REASONS FOR FAILURE	REMARKS	ACTION	✓
1. Any relative movement between steering wheel and steering column	⚠ ⚠	Tighten/Replace	
2. Any abnormal movement indicating column top bearing is excessively worn	⚠	Replace	
3. Any movement up and down in the centre of steering column due to ▪ worn steering box ▪ deteriorated flexible coupling ▪ worn upper column bearings	⚠ ⚠	Tighten/Repair Replace	
4. Any movement due to the top mounting bracket being ▪ insecure, or ▪ fractured	⚠ ⚠ ⚠	Tighten/Repair Replace	
5. The spoke, hub or rim of steering wheel is ▪ loose or ▪ fractured	⚠ ⚠ ⚠	Repair/Replace	

WARNING

Extremely dangerous. DO NOT drive your car in this condition.
You will be breaking the law and risking your life and the lives of others.

Very dangerous and may be illegal. Put right immediately.

Could also be dangerous and illegal.

Checks outside car

Is your car's steering safe?

WHAT THE MOT REQUIRES	HOW TO CHECK
The steering system The condition of the STEERING must not cause danger to any person in the car or on the road The STEERING must be in good and efficient working order and be correctly adjusted	Put the wheels in a straight ahead position supporting the weight of the car a. Check there is no excess movement of steering wheel when drop arm or steering rod is felt or seen to move slightly b. Get a friend to turn steering wheel from side to side, while you check for Movement between ▪ chassis and steering box or rack ▪ connecting joints in the mechanical linkage Insecure ▪ parts of steering box or rack ▪ balljoints ▪ all nuts and locking devices ▪ attachment of steering damper Damage, corrosion or fracture of ▪ area around steering box, rack or idler ▪ steering damper body or cover Excess ▪ fluid leakage from a steering damper ▪ wear in pivot point, ball joint and any other joints The condition of drive shaft universal joint couplings (front wheel drive cars only) **Rear wheel steering** If applicable, the above checks should also be carried out on rear wheel steering components including connecting rod.

Is your car's steering mechanism in good condition?

MAIN REASONS FOR FAILURE	REMARKS	ACTION	✓
1. Steering wheel ▪ stiff ▪ too much free play	⚠ ⚠ ⚠	Adjust/Replace	
2. Too much movement between mechanical linkage and connecting joints	⚠ ⚠ ⚠	Tighten/Replace	
3. Insecure ▪ track rod end ▪ drag link end ▪ ball pin shank ▪ steering box/rack ▪ pinion assembly ▪ bolt ▪ retaining device ▪ locking device ▪ steering damper ▪ drive shaft constant velocity joint ▪ universal joint coupling ▪ U bolt securing a joint bearing ▪ rear wheel steering connector rod	⚠ ⚠ ⚠	Tighten/Replace	
4. Damaged, corroded, or fractured ▪ bushing material ▪ flexible rubber ▪ steering component ▪ sector shaft ▪ damper body ▪ steering box/rack ▪ pinion housing bolt ▪ fabric universal coupling unit	⚠ ⚠ ⚠	Repair/Replace	
5. Fluid leakage from a steering damper indicating failure	⚠ ⚠	Repair/Replace	
6. The loadbearing member and/or panel within 30cm of steering box, rack and pinion housing, or mounting point of any part of steering assembly is ▪ excessively corroded ▪ deformed, or ▪ fractured	⚠ ⚠	Repair/Replace	
7. Hydraulic fluid leaking from a rear wheel steering system	⚠	Repair/Replace	

For interpretation of Remarks column, see page 21.

STEERING - 5

Is your car's steering safe?

WHAT THE MOT REQUIRES	HOW TO CHECK
Power steering The condition of the STEERING must not cause danger to any person in the car or on the road The STEERING must be in good and efficient working order and be correctly adjusted	Select neutral gear. Put parking brake on and start engine. Turn steering wheel from side to side and check a. That all parts similar to normal steering mechanism are correct b. That no hydraulic fluid hose or union is leaking (Switch off engine) c. The condition of power steering pump drive and security of pump mounting d. For power assistance

Is your car's power steering in good condition?

MAIN REASONS FOR FAILURE	REMARKS	ACTION	✓
1. Power steering does not operate correctly	⚠⚠⚠	Repair/Replace	
2. Any part is insecure	⚠⚠⚠	Tighten/Replace	
3. Excessive play in a power steering mechanism joint	⚠⚠	Tighten/Replace	
4. Excessive deterioration of any bushing material	⚠⚠	Replace	
5. Fluid leak or damaged hose	⚠⚠⚠	Repair/Replace	

WARNING

Extremely dangerous. DO NOT drive your car in this condition.
You will be breaking the law and risking your life and the lives of others.

Very dangerous and may be illegal. Put right immediately.

SUSPENSION & WHEEL BEARINGS - 1

Are your car's front and rear wheel bearings in good condition?

WHAT THE MOT REQUIRES	HOW TO CHECK
The wheel bearings The condition of the BEARINGS must not cause danger to any person in the car or on the road The BEARINGS must be maintained in good and efficient working order	a. Preferably with the car over a pit or on a raised hoist, jack up front and rear wheels in turn (for jacking positions, see pages 30-34) b. While front wheels are being jacked up, look for movement in inner wishbone bearings c. When car is jacked up, check for ▪ roughness in bearings when spinning the wheel ▪ tightness or excessive play when the wheel is stationary. Do this with wheel in at least two positions d. For front wheel drive cars, ▪ put gear in neutral ▪ spin wheels at each steering lock and check - condition of gaiters, and - front wheel drive shafts for straightness and damage

SUSPENSION & WHEEL BEARINGS - 2

Are your car's drive shafts in good condition (FWD only)?

MAIN REASONS FOR FAILURE	REMARKS	ACTION	✓
1. A wishbone bearing or pin ▪ worn ▪ corroded ▪ bent	⚠	Replace	
2. A flexible rubber bush seriously deteriorated	⚠	Replace	
3. A wheel bearing ▪ rough ▪ tight ▪ has excessive play, or ▪ has not enough clearance	⚠	Adjust/Replace	
4. A universal joint gaiter ▪ split ▪ missing ▪ insecurely mounted	⚠ ⚠	Adjust/Replace	
5. A damaged or bent ▪ shaft ▪ locking device	⚠ ⚠	Repair/Replace	
6. A locking device is missing or defective	⚠ ⚠	Repair/Replace	

WARNING

Very dangerous and may be illegal. Put right immediately.

Could also be dangerous and illegal.

SUSPENSION - 1

Is your car's suspension in good condition?

WHAT THE MOT REQUIRES	HOW TO CHECK
All suspension types The condition of the SUSPENSION must not cause danger to any person in the car or on the road The SUSPENSION must be in good and efficient working order and be correctly adjusted	a. Place front wheels on a device or a surface which enables wheels to turn without much resistance b. While turning wheels from lock to lock, check ▪ the flexible hoses are not fouling any moving part ▪ lockstops for security and correct adjustment ▪ condition and security of steering rack gaiters ▪ there is no tightness or roughness c. Examine condition of ▪ wishbones ▪ inner bearings ▪ track control arms ▪ suspension radius rods ▪ mounting bushes ▪ thrust washers or bearings ▪ suspension spring d. Correct mounting of ▪ suspension spring ▪ displacer ▪ bellows

SUSPENSION - 2

Is your car's steering mechanism in good condition?

MAIN REASONS FOR FAILURE	REMARKS	ACTION	✓
1. Any part of the steering mechanism ▪ interfering with fixed part of car ▪ too tight ▪ insecure ▪ rough in operation	⚠⚠	Adjust/Replace	
2. A lock stop ▪ incorrectly adjusted ▪ insecure ▪ loose ▪ damaged	⚠⚠	Adjust/Replace	
3. A steering rack gaiter ▪ insecure ▪ split ▪ missing	⚠⚠	Replace	
4. A metal or flexible brake hose is ▪ stretched ▪ twisted ▪ seriously damaged by fouling	⚠⚠⚠	Replace	
5. Excessive wear in a ▪ pin ▪ wishbone bearing or seriously deteriorated flexible bush	⚠⚠	Replace	
6. A wishbone, track control arm, or radius rod is fractured, or ▪ distorted ▪ insecure ▪ excessively corroded	⚠⚠⚠ ⚠⚠	Replace	
7. A seriously deteriorated ▪ radius arm ▪ flexible bearing ▪ thrust washer or bearing	⚠⚠	Replace	
8. A radius arm nut insecurely locked	⚠⚠	Lock/Replace	

WARNING

Extremely dangerous. DO NOT drive your car in this condition.
You will be breaking the law and risking your life and the lives of others.

Very dangerous and may be illegal. Put right immediately.

SUSPENSION - 3

Is your car's suspension in good condition?

WHAT THE MOT REQUIRES

Suspension types 1 & 2
(See Figures 1 & 2 below)

The condition of the SUSPENSION must not cause danger to any person in the car or on the road

The SUSPENSION must be in good and efficient working order and be correctly adjusted

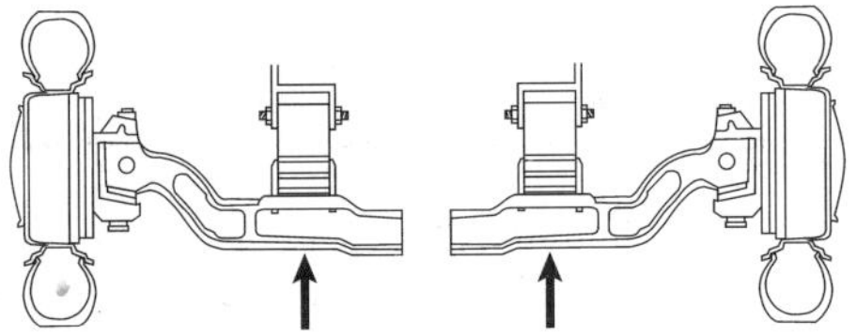

Figure 1. Suspension type 1
Jack up at points indicated by arrows*

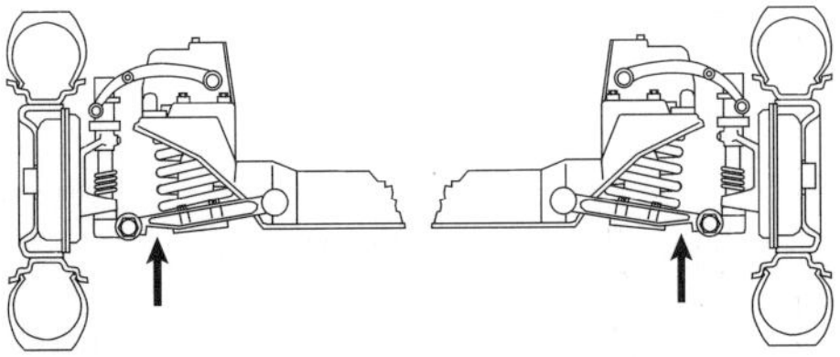

Figure 2. Suspension type 2
Jack up at points indicated by arrows*

*Use suitable jacks. Your car jack might not be the right type for this job.

HOW TO CHECK

a. Jack up front suspension so that wheels are clear of the ground

b. Ask a friend to
   - hold wheel at top and bottom (see Figure 5 on page 33), and
   - rock wheel backwards and forwards

Check for movement

- between kingpin and its bushes or in axle boss
- between wishbone outer suspension ball joints and their housings
- in upper inner wishbone bearings

c. With a bar under each front roadwheel, try to lift wheel and check for movement between
   - stub axle yoke and its housing at the thrust bearing
   - suspension ball joints and their housing

d. Check beam axles, wishbones and stub axles for damage and distortion

e. Examine condition of chassis frame and body shell structure around suspension mounting points for fractures, corrosion and distortion

SUSPENSION - 4

Is your car's steering mechanism in good condition?

MAIN REASONS FOR FAILURE	REMARKS	ACTION	✓
1. A kingpin is loose in its ▪ mounting boss or ▪ stub axle bush	⚠⚠⚠	Tighten/Replace	
2. A kingpin retaining device is loose or missing	⚠⚠⚠	Tighten/Replace	
3. Excessive wear in ▪ suspension swivel pin ▪ suspension ball joint ▪ wishbone bearing or flexible bush	⚠⚠	Replace	
4. Excessive lift between stub axle and axle housing	⚠⚠	Adjust/Repair	
5. Excessive play caused by wear in a ball joint	⚠⚠	Replace	
6. A distorted axle beam or component	⚠⚠	Replace	
7. A wishbone arm is fractured or ▪ excessively corroded ▪ distorted	⚠⚠⚠ ⚠⚠	Replace	
8. The loadbearing member and/or panel within 30cm of the mounting point for the suspension is ▪ excessively corroded ▪ deformed ▪ fractured	⚠⚠	Repair/Replace	

WARNING

Extremely dangerous. DO NOT drive your car in this condition.
You will be breaking the law and risking your life and the lives of others.

Very dangerous and may be illegal. Put right immediately.

SUSPENSION - 5

Is your car's suspension in good condition?

WHAT THE MOT REQUIRES	HOW TO CHECK
Suspension types 3 & 3a (See Figures 3 & 4 below) The condition of the SUSPENSION must not cause danger to any person in the car or on the road The SUSPENSION must be in good and efficient working order and be correctly adjusted	a. Place the front wheels on a device or a surface which enables the wheels to turn without much resistance b. Shake wheel vigorously (see Figures 6 & 7 on opposite page) and check for ▪ play between ball and its housing in suspension ball joint ▪ a seriously worn pin or bush in an inner wishbone bearing

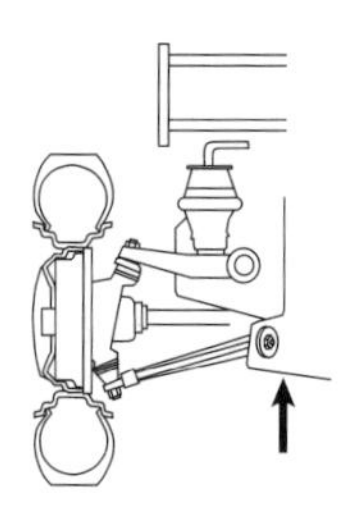
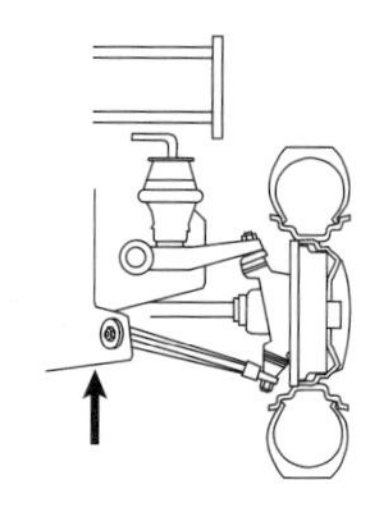

Figure 3. Suspension type 3
Jack up at points indicated by arrows*

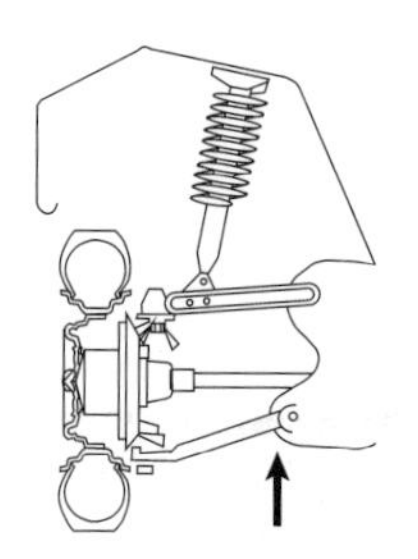
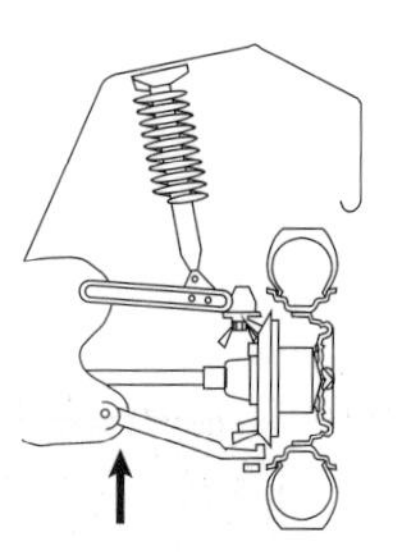

Figure 4. Suspension type 3a
Jack up at points indicated by arrows*

*Use suitable jacks. Your car jack might not be the right type for this job.

Is your car's suspension safe?

MAIN REASONS FOR FAILURE	REMARKS	ACTION	✓
1. Suspension ball joint seriously worn	⚠ ⚠ ⚠	Replace	
2. A ball joint nut/securing nut is ▪ loose ▪ not locked	⚠ ⚠	Tighten/Lock	
3. A seriously worn ▪ pin ▪ bush ▪ inner wishbone bearing	⚠ ⚠ ⚠	Replace	

WARNING

Extremely dangerous. DO NOT drive your car in this condition.
You will be breaking the law and risking your life and the lives of others.

Very dangerous and may be illegal. Put right immediately.

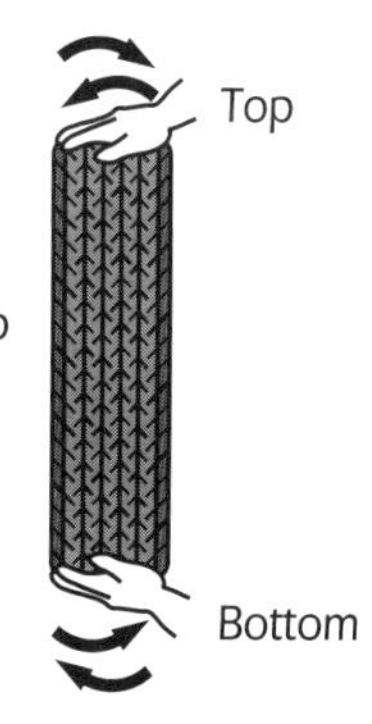

Figure 5.

Hands at 6 o'clock and 12 o'clock positions (top and bottom of wheel) Rock wheel backwards and forwards.

Figure 6.

Both hands at 12 o'clock position (top of wheel). Push and pull wheel.

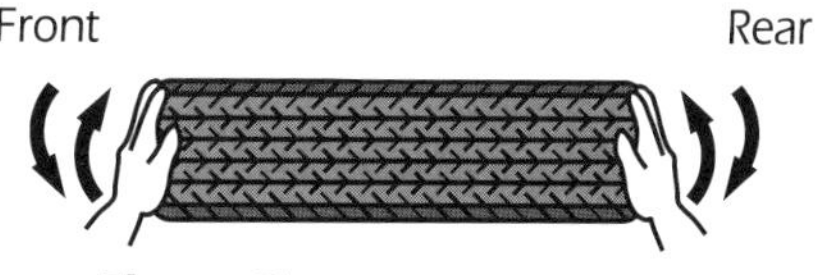

Figure 7.

Hands at 3 o'clock and 9 o'clock positions (each side of wheel). Swivel wheel in short, sharp steering actions.

SUSPENSION - 7

Is your car's suspension in good condition?

WHAT THE MOT REQUIRES

Suspension type 4
(See Figure 8 below)

The condition of the SUSPENSION must not cause danger to any person in the car or on the road

The SUSPENSION must be in good and efficient working order and be correctly adjusted

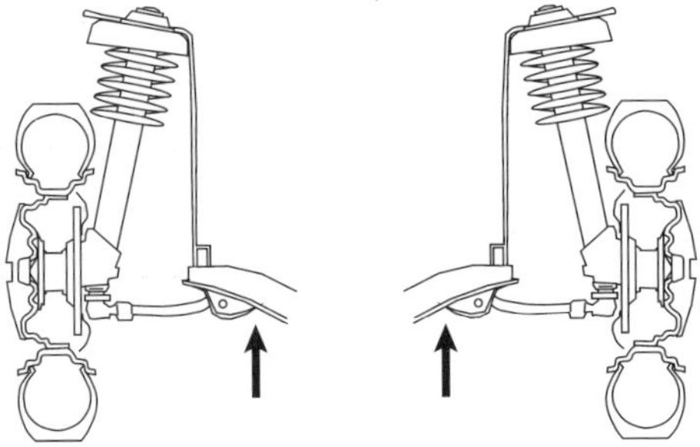

Figure 8. Suspension type 4
Jack up at points indicated by arrows*

*Use suitable jacks. Your car jack might not be the right type for this job.

HOW TO CHECK

a. Place front wheels on a device or surface which enables wheels to turn without much resistance

b. Place both hands on top of wheel and push and pull wheel
(See Figure 6 on page 33)

c. Check for

- wear in a shock absorber strut and/or bush
- wear in a rod location
- movement at upper support bearing
- leak of fluid from gland
- corrosion or damage to strut casing
- condition of bonding between metal and flexible material in strut upper-support bearing

d. Hold wheel at 3 o'clock and 9 o'clock positions (see Figure7 on page 33). Swivel wheel in short, sharp steering actions and check for movement of

- strut lower ball joint
- track control arm inner bushes

e. Check each road spring coil for

- correct mounting
- damage

SUSPENSION - 8

Is your car's suspension secure?

MAIN REASONS FOR FAILURE	REMARKS	ACTION	✓
1. A suspension strut and/or bush is worn	⚠ ⚠	Replace	
2. Shock absorber gland leaking fluid	⚠ ⚠	Repair/Replace	
3. A strut upper bearing assembly is ▪ rough ▪ stiff ▪ has too much free play	⚠ ⚠	Adjust/Replace	
4. A strut casing has ▪ damage ▪ excessive corrosion	⚠ ⚠	Replace	
5. Excessive deterioration in bonding between metal and flexible material of an upper support bearing	⚠ ⚠	Replace	
6. A lock nut in upper support bearing ▪ loose ▪ insecurely locked	⚠ ⚠	Tighten/Lock	
7. Excessive deterioration in bonding or flexible material of ▪ track control arm ▪ radius member bush ▪ radius member mounting	⚠ ⚠	Replace	
8. Excessive wear in a strut lower ball joint	⚠ ⚠ ⚠	Repair/Replace	
9. A ball joint or assembly cover nut is ▪ loose ▪ insecurely locked	⚠ ⚠ ⚠	Tighten/Lock	
10. The loadbearing member and/or panel within 30cm of the mounting point for the suspension is ▪ excessively corroded ▪ deformed ▪ fractured	⚠ ⚠	Repair/Replace	

WARNING

Extremely dangerous. DO NOT drive your car in this condition.
You will be breaking the law and risking your life and the lives of others.

Very dangerous and may be illegal. Put right immediately.

SUSPENSION ASSEMBLIES - 1

Is your car's sub-frame secure?

WHAT THE MOT REQUIRES	HOW TO CHECK
The sub-frame The condition of the SUB-FRAME must not cause danger to any person in the car or on the road	a. Examine the condition of sub-frame b. Check mountings for ▪ fractures ▪ excessive corrosion ▪ insecurity ▪ deterioration

SUSPENSION ASSEMBLIES - 2

Is your car's sub-frame in good condition?

MAIN REASONS FOR FAILURE	REMARKS	ACTION	✓
1. A sub-frame is badly ▪ distorted ▪ fractured ▪ corroded ▪ repaired	 ⚠ ⚠ ⚠ ⚠ ⚠ ⚠ ⚠ ⚠ ⚠	Repair/Replace	
2. An insecurely locked or defective mounting	⚠ ⚠	Lock/Replace	
3. A badly deteriorated flexible mounting	⚠	Replace	
4. The loadbearing member and/or panel within 30cm of the mounting point for the sub-frame is ▪ excessively corroded ▪ deformed ▪ fractured	⚠ ⚠	Repair/Replace	

WARNING

Extremely dangerous. DO NOT drive your car in this condition.
You will be breaking the law and risking your life and the lives of others.

Very dangerous and may be illegal. Put right immediately.

Could also be dangerous and illegal.

SUSPENSION ASSEMBLIES - 3

Is your car's suspension assembly in good condition?

WHAT THE MOT REQUIRES	HOW TO CHECK
The coil spring or displacer units The condition of the ASSEMBLY must not cause danger to any person in the car or on the road The ASSEMBLY must be in good and efficient working order	**How to check the condition of your coil spring and displacer units** a. Check each coil spring and displacer unit for ▪ correct mounting ▪ damage b. Check car's structure near mounting point for ▪ fractures ▪ excessive corrosion ▪ distortion c. Examine any interconnecting pipes between displacer units

SUSPENSION ASSEMBLIES - 4

Is your car's suspension assembly secure?

MAIN REASONS FOR FAILURE	REMARKS	ACTION	✓
1. A coil spring ▪ incomplete ▪ fractured ▪ cross section seriously reduced due to corrosion or wear	⚠⚠⚠	Replace	
2. A coil spring or displacer unit incorrectly seated	⚠⚠⚠	Adjust	
3. Inadequate clearance of the axle or suspension with the bump stop or chassis	⚠	Repair/Replace	
4. Interconnecting pipes between displacer units are ▪ damaged ▪ excessively corroded ▪ insecure ▪ leaking	⚠⚠	Replace	
5. The loadbearing member and/or panel within 30cm of the mounting point for the spring is ▪ excessively corroded ▪ deformed ▪ fractured	⚠⚠	Repair/Replace	

WARNING

Extremely dangerous. DO NOT drive your car in this condition.
You will be breaking the law and risking your life and the lives of others.

Very dangerous and may be illegal. Put right immediately.

Could also be dangerous and illegal.

SUSPENSION ASSEMBLIES - 5

Is your car's suspension assembly in good condition?

WHAT THE MOT REQUIRES	HOW TO CHECK
The leaf spring The condition of the ASSEMBLY must not cause danger to any person in the car or on the road The ASSEMBLY must be in good and efficient working order	a. Check each leaf spring for ▪ fractures ▪ displaced leaves ▪ deformation b. Check that each spring is fitted so that ▪ the axle is correctly located ▪ it is secured to the axle c. Check ▪ spring anchor bracket ▪ spring shackle bracket ▪ associated pins and bushes for ▪ wear ▪ security ▪ adequate locking ▪ side play d. Check car's structure within 30cm of any spring mounting for ▪ fractures ▪ excessive corrosion ▪ distortion

SUSPENSION ASSEMBLIES - 6

Is your car's suspension assembly secure?

MAIN REASONS FOR FAILURE	REMARKS	ACTION	✓
1. A leaf in a spring is ▪ incomplete ▪ fractured ▪ splayed excessively	⚠⚠⚠	Repair/Replace	
2. Not enough clearance of the axle or suspension with the bump stop or chassis	⚠	Repair/Replace	
3. A broken centre bolt	⚠⚠⚠	Replace	
4. Spring incorrectly fitted so the axle is incorrectly located	⚠⚠⚠	Adjust/Replace	
5. A loose securing bolt or plate	⚠⚠	Adjust/Replace	
6. A defective spring eye	⚠⚠	Replace	
7. An anchor pin and bush ▪ loose in its bracket ▪ insecure ▪ excessively worn ▪ missing	⚠⚠⚠	Adjust/Replace	
8. A shackle pin and/or bush ▪ loose in its bracket ▪ insecure ▪ excessively worn ▪ missing	⚠⚠	Adjust/Replace	
9. The loadbearing member and/or panel within 30cm of the mounting point for the spring is ▪ excessively corroded ▪ deformed ▪ fractured	⚠⚠⚠	Repair/Replace	

WARNING

Extremely dangerous. DO NOT drive your car in this condition.
You will be breaking the law and risking your life and the lives of others.

Very dangerous and may be illegal. Put right immediately.

Could also be dangerous and illegal.

SUSPENSION ASSEMBLIES - 7

Is your car's suspension assembly in good condition?

WHAT THE MOT REQUIRES	HOW TO CHECK
Radius arms, links and tie bars The condition of the ASSEMBLY must not cause danger to any person in the car or on the road The ASSEMBLY must be in good and efficient working order and be correctly adjusted	**How to check the condition of radius arms, links and tie bars.** a. Check each radius arm, link and tie bar for ▪ fractures ▪ distortion ▪ excessive corrosion b. Check mountings for ▪ wear ▪ security c. On vehicles which have a drive shaft which forms part of suspension, check for ▪ distortion ▪ damage ▪ serious corrosion On these vehicles check universal joint bearings for wear, and flanges and bolts for security d. Check car's structure and panel within 30cm of arm/ link mountings for ▪ fractures ▪ excessive corrosion ▪ distortion

SUSPENSION ASSEMBLIES - 8

Is your car's suspension assembly secure?

MAIN REASONS FOR FAILURE	REMARKS	ACTION	✓
1. A radius arm or locating link ▪ distorted ▪ fractured ▪ excessively corroded	⚠⚠⚠	Replace	
2. Excessive wear in a radius arm or locating link bush or bearing	⚠⚠	Replace	
3. A radius arm or locating link mounting or pin ▪ insecure ▪ not properly locked	⚠⚠	Tighten/Lock	
4. Serious deterioration of bonding or flexible material of a radius arm or link	⚠⚠	Repair/Replace	
5. A drive shaft ▪ distorted ▪ damaged ▪ excessively corroded	⚠⚠	Replace	
6. Excessively worn universal joint bearing	⚠⚠	Replace	
7. An incorrectly seated universal joint flange	⚠⚠	Adjust/Replace	
8. A loose flange bolt	⚠⚠	Tighten/Lock	
9. The loadbearing member and/or panel within 30cm of the mounting point for radius arm or locating link is ▪ excessively corroded ▪ deformed ▪ fractured	⚠⚠⚠	Repair/Replace	

WARNING

Extremely dangerous. DO NOT drive your car in this condition.
You will be breaking the law and risking your life and the lives of others.

Very dangerous and may be illegal. Put right immediately.

SUSPENSION ASSEMBLIES - 9

Is your car's suspension assembly in good condition?

WHAT THE MOT REQUIRES	HOW TO CHECK
Torsion bars The condition of the ASSEMBLY must not cause danger to any person in the car or on the road The ASSEMBLY must be in good and efficient working order and be correctly adjusted	a. Check each torsion bar for ▪ fractures ▪ distortion ▪ excessive corrosion ▪ pitting ▪ free play where torsion bar is connected to suspension arm, guide or wishbone b. Check security of torsion bar abutment screw assembly to the body structure of the car c. Check the car's structure within 30cm of the mounting point of a torsion bar attachment for ▪ fractures ▪ excessive corrosion ▪ distortion

SUSPENSION ASSEMBLIES - 10

Is your car's suspension assembly secure?

MAIN REASONS FOR FAILURE	REMARKS	ACTION	✓
1. A torsion bar ▪ distorted ▪ fractured ▪ excessively corroded	⚠⚠⚠	Replace	
2. Excessive free play where torsion bar is connected to suspension arm or wishbone	⚠⚠	Replace	
3. A torsion bar abutment ▪ damaged ▪ not properly locked	⚠⚠	Tighten/Replace	
4. Not enough clearance of axle or suspension with bump stop or chassis	⚠	Adjust/Replace	
5. The loadbearing member and/or panel within 30cm of the mounting point for a torsion bar attachment is ▪ excessively corroded ▪ deformed ▪ fractured	⚠⚠⚠	Repair/Replace	

WARNING

Extremely dangerous. DO NOT drive your car in this condition.
You will be breaking the law and risking your life and the lives of others.

Very dangerous and may be illegal. Put right immediately.

Could also be dangerous and illegal.

SUSPENSION ASSEMBLIES - 11

Is your car's suspension assembly in good condition?

WHAT THE MOT REQUIRES	HOW TO CHECK
Anti-roll bar The condition of the ASSEMBLY must not cause danger to any person in the car or on the road The ASSEMBLY must be in good and efficient working order and be correctly adjusted	a. Check each anti-roll bar for ▪ fractures ▪ distortion ▪ excessive corrosion ▪ security ▪ wear in the bearings or joints b. Check car's structure within 30cm of the mounting point for an anti-roll bar for ▪ fractures ▪ excessive corrosion ▪ distortion

SUSPENSION ASSEMBLIES - 12

Is your car's suspension assembly secure?

MAIN REASONS FOR FAILURE	REMARKS	ACTION	✓
1. An anti-roll bar ▪ fractured ▪ distorted ▪ excessively corroded	⚠ ⚠ ⚠	Replace	
2. A serious deterioration of bonding or flexible material of an anti-roll bar joint or bearing	⚠ ⚠	Repair/Replace	
3. The loadbearing member and/or panel within 30cm of the mounting point for an anti-roll bar is ▪ excessively corroded ▪ deformed ▪ fractured	⚠ ⚠ ⚠	Repair/Replace	

WARNING

Extremely dangerous. DO NOT drive your car in this condition.
You will be breaking the law and risking your life and the lives of others.

Very dangerous and may be illegal. Put right immediately.

SHOCK ABSORBERS - 1

Are your shock absorbers working properly?

WHAT THE MOT REQUIRES	HOW TO CHECK
The shock absorbers The condition of the SHOCK ABSORBERS must not cause danger to any person in the car or on the road The SHOCK ABSORBERS must be in good and efficient working order and be correctly adjusted	a. Check for presence of shock absorbers b. Check each shock absorber for ▪ damage ▪ fluid leaks ▪ insecurity **Note 1** Make sure any leak is from the unit and not from another source **Note 2** Where flexible dust shields are fitted, squeeze shield and watch for a fluid leak under it c. Push down and release each corner of the car. Check if shock absorbers are damping effectively **Note** A rough measure of effective damping is if the corner of the car bounces not more than one and a half times when pushed down and released

SHOCK ABSORBERS - 2

Are your shock absorbers secure?

MAIN REASONS FOR FAILURE	REMARKS	ACTION	✓
1. Extensive fluid leakage indicating failed seal	⚠⚠	Replace	
2. A shock absorber ▪ badly damaged ▪ corroded ▪ insecure ▪ missing ▪ not producing damping effect	⚠⚠	Tighten/Adjust Replace	
3. A shock absorber lever or link is insecurely attached	⚠⚠	Tighten	

WARNING

Very dangerous and may be illegal. Put right immediately.

BRAKES - 1

Does your parking brake work properly?

WHAT THE MOT REQUIRES	HOW TO CHECK
The brake controls **The parking brake** The condition of the PARKING BRAKE must not cause danger to any person in the car or on the road The PARKING BRAKE must be mechanical, in good and efficient working order and be correctly adjusted	a. While parking brake lever is being operated, check that at least two wheels are prevented from turning b. Move lever sideways to check for wear in pivot bearing c. Apply parking brake and check that it can be set in the 'on' position and will stay 'on' even when tapped d. Check mountings, structure and panelling for ▪ security ▪ condition

BRAKES - 2

Is your parking brake in good working order?

MAIN REASONS FOR FAILURE	REMARKS	ACTION	✓
1. The car does not have required parking brake system	⚠⚠⚠	Replace	
2. With reasonable force, lever cannot be set in 'on' position	⚠⚠⚠	Repair/Replace	
3. Lever releases with slight sideways movement or accidental contact	⚠⚠⚠	Repair/Replace	
4. Parking brake mechanism not securely attached to car	⚠⚠⚠	Repair/Replace	
5. Loadbearing member and/or panel within 30cm of mounting point for handbrake is ▪ corroded ▪ fractured ▪ distorted	⚠⚠⚠	Repair/Replace	
6. The amount of pull before parking brake comes 'on' is more than normal for that type	⚠	Check wear Adjust	

WARNING

Extremely dangerous. DO NOT drive your car in this condition.
You will be breaking the law and risking your life and the lives of others.

Could also be dangerous and illegal.

BRAKES - 3

Is your footbrake working properly?

WHAT THE MOT REQUIRES	HOW TO CHECK
The brake controls **The footbrake** The condition of the FOOTBRAKE must not cause danger to any person in the car or on the road The FOOTBRAKE must be in good and efficient working order and be correctly adjusted	From the driver's seat a. Check ▪ wear at pedal/pedal pivot ▪ pedal is complete b. Press pedal slowly until pressure can be held Check for creep* Press pedal quickly until pressure can be held Check again for creep c. Check for fluid leaks

*'Creep': under constant pressure, a pedal moves slowly indicating an internal or external leak in the system.

Is your footbrake in good working order?

MAIN REASONS FOR FAILURE	REMARKS	ACTION	✓
1. The footbrake does not work	⚠⚠⚠	Repair/Replace	
2. Pedal has inadequate travel reserve	⚠⚠⚠	Check wear Adjust	
3. With sustained pressure, pedal creeps	⚠⚠⚠	Repair/Replace	
4. Pedal movement is spongy*	⚠⚠⚠	Replace	
5. Obvious leak in master cylinder	⚠⚠⚠	Repair/Replace	

WARNING

Extremely dangerous. DO NOT drive your car in this condition.
You will be breaking the law and risking your life and the lives of others.

*'Spongy': easy to compress (not hard or solid). If there is air in the system, the brakes will feel 'spongy' when applied, and there may be too much travel (distance between the resting position of the brake pedal and the point where the brakes apply when the pedal is pressed).

BRAKES - 5

Does the brake system work properly?

WHAT THE MOT REQUIRES	HOW TO CHECK
The braking system The condition of the BRAKES must not cause danger to any person in the car or on the road The BRAKES must be in good and efficient working order and be correctly adjusted	**Parking brake mechanism (under vehicle)** a. Examine all the mechanical parts you can see, without dismantling. Get a friend to keep putting brakes on and off while you check for ▪ corrosion ▪ insecurity ▪ cracking ▪ restrictions to free movement ▪ abnormal movement ▪ too much wear ▪ incorrect adjustment ▪ secure holding on mechanism b. Check ▪ lever pivots ▪ cables ▪ rods ▪ linkages ▪ outer casings ▪ guides ▪ clevis pins ▪ yokes ▪ locking devices

Is the parking brake mechanism in good working order?

MAIN REASONS FOR FAILURE	REMARKS	ACTION	✓
1. Serious weakening of any component due to ▪ cracking ▪ too much wear ▪ damage	⚠⚠⚠	Replace	
2. A corroded cable	⚠	Replace	
3. A knotted cable	⚠⚠	Replace	
4. A frayed cable	⚠⚠⚠	Replace	
5. A cable outer casing is damaged	⚠⚠	Repair/Replace	
6. Holding-on mechanism is badly worn	⚠⚠⚠	Repair/Replace	
7. A parking brake lever attachment is insecure	⚠⚠	Tighten Repair/Replace	
8. Any restriction to free movement of system	⚠⚠	Repair	
9. Unusual movement indicating bad adjustment or too much wear	⚠⚠	Adjust/Replace	
10. The loadbearing member and/or panel within 30cm of the mounting point for the parking brake lever attachment is ▪ excessively corroded ▪ deformed ▪ fractured	⚠⚠⚠	Repair/Replace	

WARNING

Extremely dangerous. DO NOT drive your car in this condition.
You will be breaking the law and risking your life and the lives of others.

Very dangerous and may be illegal. Put right immediately.

Could also be dangerous and illegal.

BRAKES - 7

Does the brake system work properly?

WHAT THE MOT REQUIRES	HOW TO CHECK
The footbrake system The condition of the BRAKES must not cause danger to any person in the car or on the road The BRAKES must be in good and efficient working order and be correctly adjusted Cars registered on or after 1 January 1915 must have ▪ two efficient braking systems, with separate controls, or ▪ one system with two separate controls When one brake is applied, it must not effect the working of the other Cars registered before 1 Jan 1915 and three wheeled vehicles must have at least one braking system	**Hydraulic brake components** a. Check hydraulic reservoirs and cylinders for ▪ security of mounting ▪ damage ▪ leaks b. Check all visible brake pipes and flexible hoses for ▪ chafing ▪ corrosion ▪ damage ▪ security ▪ fouling ▪ leaks ▪ kinking ▪ stretching ▪ repairs c. Check for ▪ excessive corrosion at master cylinder mounting ▪ insecure brake back plate or disc caliper housing ▪ vacuum servo not working (see Note) d. Get a friend to apply brakes firmly while you check for ▪ leaks ▪ bulges in hoses **Note** To check vacuum servo 1. With engine switched off, press brake pedal several times to deplete all stored vacuum in servo 2. Hold brake on firmly with left foot and start engine 3. If servo is working, brake pedal will travel further as vacuum builds up in system

BRAKES - 8

Is the footbrake system in good working order?

MAIN REASONS FOR FAILURE	REMARKS	ACTION	✓
1. A reservoir or master cylinder ▪ insecurely mounted ▪ damaged	⚠ ⚠ ⚠	Repair/Replace	
2. Leak of hydraulic fluid	⚠ ⚠ ⚠	Repair/Replace	
3. Pipes ▪ chafed ▪ kinked ▪ damaged ▪ corroded ▪ cracked ▪ inadequately supported/repaired ▪ likely to hinder or be trapped by moving parts	⚠ ⚠ ⚠	Replace	
4. Hoses ▪ twisted ▪ damaged ▪ chafed ▪ deteriorated ▪ cracked ▪ bulging under pressure ▪ stretched by steering or suspension ▪ likely to hinder or be trapped by moving parts	⚠ ⚠ ⚠	Replace	
5. Vacuum servo not working	⚠ ⚠	Repair/Replace	
6. Insecure or corroded brake back plate or disc caliper housing	⚠ ⚠ ⚠	Repair	
7. A disc or drum is fractured, or ▪ excessively pitted ▪ scored ▪ worn	⚠ ⚠ ⚠ ⚠ ⚠	Replace Replair/Replace	
8. The loadbearing member and/or panel within 30cm of the mounting point for the master cylinder is ▪ corroded ▪ fractured ▪ distorted	⚠ ⚠ ⚠	Repair/Replace	

WARNING

Extremely dangerous. DO NOT drive your car in this condition.
You will be breaking the law and risking your life and the lives of others.

Very dangerous and may be illegal. Put right immediately.

BRAKES - 9

Does the brake system work properly?

WHAT THE MOT REQUIRES	HOW TO CHECK
Anti-lock system The ABS inspection applies to all systems fitted as 'standard' and to 'optional' systems fitted.	If the vehicle is fitted with an anti-lock braking system, check that the warning lamp a. illuminates b. follows the correct sequence of operation **Note:** The sequence varies with the type of system. Refer to the manufacturer's or other reliable data.

Does the Anti-lock braking system (ABS) work properly?

MAIN REASONS FOR FAILURE	REMARKS	ACTION	✓
1. Warning lamp ▪ does not illuminate ▪ does not follow correct sequence ▪ indicates an ABS fault	⚠	Repair	

WARNING

Could also be dangerous and illegal.

BRAKES - 11

Are your brakes working correctly?

WHAT THE MOT REQUIRES	HOW TO CHECK

The braking systems

The condition of the BRAKES must not cause danger to any person in the car or on the road

The BRAKES must be in good and efficient working order and be correctly adjusted

Brake Efficiency Table	
Effiicency requirement (see page 63)	Metres from marker to car front
50%	8
40%	10
30%	13
25%	16
16%	25

Note. The 'MOT' brake test is more comprehensive. Make sure your brakes are working correctly.

Brake performance test

1. The road
   - Choose a quiet, level road
   - Avoid residential streets or causing nuisance
   - Make sure no traffic is coming in either direction
2. Drive at 20 mph
3. Press clutch pedal and apply footbrake with constantly increasing pressure

   Brakes must work
   - effectively
   - progressively
   - without juddering
   - without fluctuation
   - without the vehicle deviating or swerving
4. Repeat test using handbrake only, keeping ratchet disengaged (button pressed) all the time

To do a more accurate test

1. Select marker on road (road sign, tree, etc.) Drive at 20 mph
2. As car front passes marker, begin to apply brakes (as above)
3. When car has stopped, step out distance from marker to front of car. Count each stride as a metre. To find out your brake efficiency, see table on left.

Are your brakes efficient?

MAIN REASONS FOR FAILURE	REMARKS	ACTION	✓
1. Brake efficiency is less than required * (see table)	⚠⚠⚠	Repair/Adjust/ Replace	
2. Brake sticks or binds**	⚠⚠⚠	Repair/Adjust/ Replace	
3. Brake effort fluctuates when brakes are steadily applied	⚠⚠	Repair/Adjust/ Replace	
4. Severe grab or judder when brakes are applied	⚠⚠⚠	Repair/Adjust/ Replace	
5. Braking causes pulling or swerving to one side	⚠⚠⚠	Repair/Adjust/ Replace	

WARNING

Extremely dangerous. DO NOT drive your car in this condition.
You will be breaking the law and risking your life and the lives of others.

Very dangerous and may be illegal. Put right immediately.

* Brake efficiency can only be measured accurately with special equipment.

** 'Sticking': When the brakes remain on after brake pedal has been released
'Binding': When there is a braking effect with the brake control released

BRAKES - 13

Are your brakes working correctly?

WHAT THE MOT REQUIRES

Service Brake

Cars first used before 1 January 1915 need have only one efficient braking system.

All cars first used on or after that date must have

- an efficient braking system with two means of control, or
- two efficient braking systems with separate means of control, or
- one efficient braking system with one control, if the system is 'split'.

A split or dual system has two independent braking circuits which are applied by working a single means of control, usually the brake pedal.

To find out whether your car has a split or dual braking system, check the number of hydraulic pipes leading from the master cylinder. A split or dual system normally has two pipes, or two separate master cylinders.

Cars first used on or after 1 January 1968 must have one means of control of the braking system which acts on all wheels.

Secondary Brake

On a split braking system, the secondary brake is normally considered to be one half of that split system.

Where a split system is not employed, the secondary performance must be achieved by

- the second means of control, and/or
- the second braking system, usually the parking brake.

Parking Brake

All cars must have a braking system which can prevent at least two wheels (one for a three-wheel car) from turning when the car is stationary.

The parking brake for all cars first used on or after 1 January 1968 must be able, by mechanical means only, to prevent the car from moving on a 1 in 6.25 (16%) slope.

Brake Balance

The braking effort on one front wheel must be at least 75% of the effort on the other front wheel.

Measuring Brake Efficiency

Brake efficiency is usually measured using specialized equipment such as that used by the MOT testing station.

Do your brakes obtain the minimum requirement?

BRAKE EFFICIENCY — WHAT THE MOT REQUIRES

Class of Vehicle	Type of System	Service Brake	Parking Brake
Motor cars first used on or after 1 Jan 1968	**One** means of control applying to **all** wheels	50%	Split system — 16% Non-split system — 25%
Motor cars ▪ first used before 1 Jan 1968 ▪ having more than 3 wheels	**One** means of control applying to **at least 4 wheels**	50%	Split system — no specified efficiency — see note Non-split system — 25%
Three-wheeled motor cars first used before 1 Jan 1968	**One** means of control applying to **all 3 wheels**	40%	Split system — no specified efficiency — see note Non-split system — 25%

Class of Vehicle	Type of System		
Motor cars ▪ first used before 1 Jan 1968 ▪ not having one means of control applying to at least 4 wheels (or 3 for three-wheeled cars)	**One** braking system with **two** means of control or **Two** brake systems with **separate** means of control	30% from first means of control	25% from second means of control

Note: On vehicles first used before 1 January 1968 there is no specified brake efficiency for the parking brake unless it is also the secondary brake.

If it is solely the parking brake it is required to be able to prevent the rotation of at least two wheels in the case of a four wheeled vehicle from rotating while the vehicle is left unattended (one wheel on three wheeled vehicles).

WHEELS - 1

Are your wheels in good condition?

WHAT THE MOT REQUIRES	HOW TO CHECK
The wheels The condition of ANY WHEEL must not cause danger to any person in the car or on the road The spare wheel is not included in the MOT Test. However, it would be wise to make sure that the condition of the spare wheel complies with requirements. An MOT tester need not remove hub caps, wheel trims, etc	**Wear or damage** a. Check wheels, particularly the bead rim, for ▪ damage ▪ cracks ▪ distortion b. Check the wheels are secure c. If visible, check that securing nuts and studs are not ▪ loose ▪ missing **How to check for wear and damage** a. Raise the wheels clear of the ground. b. Rotate each wheel slowly and check ▪ Seating of tyre on wheel rim ▪ Surface of tyre, damage/repairs etc ▪ Tread wear. (Use depth gauge) ▪ Valve condition and alignment ▪ Tyre inflation

Are your wheels secure?

MAIN REASONS FOR FAILURE	REMARKS	ACTION	✓
1. A wheel ▪ cracked ▪ distorted ▪ damaged	⚠⚠⚠	Replace	
2. A badly distorted wheel bead rim	⚠⚠⚠	Replace	
3. An insecure wheel	⚠⚠⚠	Tighten/Replace	
4. Loose or missing ▪ wheel nuts ▪ studs ▪ bolts	⚠⚠⚠	Tighten/Replace	

WARNING

Extremely dangerous. DO NOT drive your car in this condition.
You will be breaking the law and risking your life and the lives of others.

TYRES - 1

Are your tyres suitable?

WHAT THE MOT REQUIRES	HOW TO CHECK
The tyres The condition of ANY TYRE must not cause danger to any person in the car or on the road You must not use your car on the road if the tyre on any road wheel is unsuitable ▪ for the use to which you are putting the car, or ▪ in relation to the tyres on the other wheels The spare wheel is not included in the MOT Test. However, you should make sure that the tyre on the spare wheel complies with the requirements.	**Unsuitable tyres include** ▪ a tyre of a different nominal size or structure from other tyre(s) on same axle ▪ special lightweight or space saving wheel and tyre fitted as a road wheel ▪ a tyre on a twin wheel is a different nominal size or structure from its twin ▪ radial-ply tyres fitted to front wheels and cross-ply or bias belted tyres fitted to the rear or ▪ bias-belted tyres fitted to the front wheels with cross-ply fitted to the rear **Tyre type is written on tyre sidewall as follows** Radial tyres ▪ The word 'Radial' appears on the sidewall ▪ The letter 'R' usually appears in the number, for example, 165 R 13 Bias-belted tyres ▪ The words 'Bias-belted' appear on the sidewall

MAIN REASONS FOR FAILURE	REMARKS	ACTION	✓
1. An unsuitable tyre	⚠⚠⚠	Replace	

WARNING

Extremely dangerous. DO NOT drive your car in this condition.
You will be breaking the law and risking your life and the lives of others.

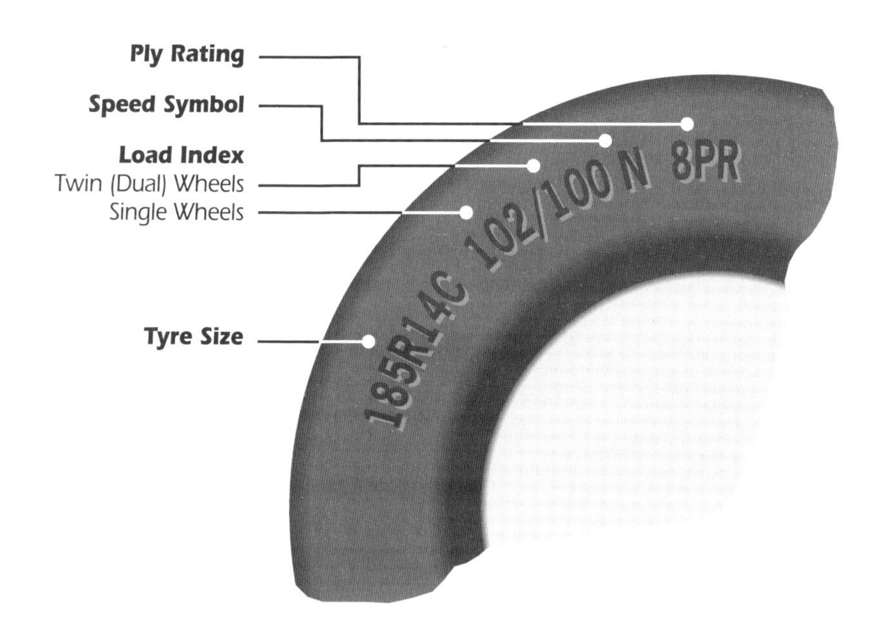

TYRES - 3

Are your tyres in good condition?

WHAT THE MOT REQUIRES	HOW TO CHECK
You must not use ANY TYRE that 1. Has a cut ▪ longer than 25mm, or 10% of the width of the tyre, whichever is greater, and/or ▪ deep enough to reach the ply or cord 2. Has a lump, bulge or tear caused by partial separation or failure of its structure 3. Has any exposed ply or cord **On a twin-wheeled vehicle,** the inner sidewalls of the tyres must not touch while the tyres are correctly inflated (See Fig. 9 below)	**Look for wear and damage** a. Raise each wheel in turn clear of the ground* b. Rotate each wheel slowly and check ▪ seating of each tyre on wheel rim ▪ surface of tyre, for damage/repairs, etc ▪ there is no evidence of fouling of other parts ▪ there is no contact between twin wheels ▪ the tyre is fitted in accordance with any specific directional markings shown on the sidewall c. Stop wheel rotating and check valve stem for ▪ cuts or other damage ▪ alignment * This check can also be carried out without raising the wheels, by moving the car to reveal the hidden parts.

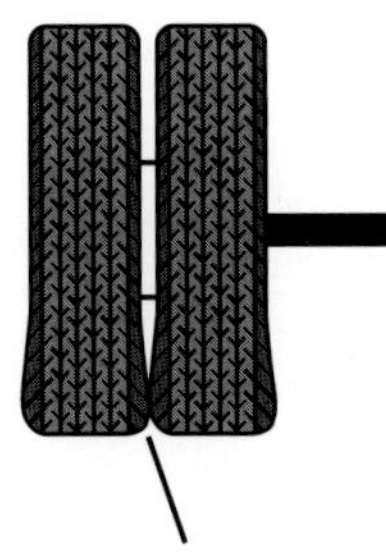

Figure 9. Touching sidewalls

If twin wheels have sidewalls which are making contact ('kissing') with the tyres correctly inflated, your vehicle will fail

Are your tyres damaged?

MAIN REASONS FOR FAILURE	REMARKS	ACTION	✓
1. A cut deep enough to reach ply or cord.	⚠ ⚠ ⚠	Replace	
2. A lump, bulge or tear caused by failure of tyre structure, such as ▪ separation ▪ tread lifting	⚠ ⚠ ⚠	Replace	
3. Any ply or cord exposed	⚠ ⚠ ⚠	Replace	
4. A valve stem ▪ out of alignment ▪ damaged	⚠ ⚠	Realign/ Replace if damaged	
5. Twin wheel side walls touching while tyre is correctly inflated	⚠ ⚠	Replace	
6. Tyre not correctly seated on wheel rim	⚠ ⚠ ⚠	Refit correctly	
7. Tyre not fitted in accordance with sidewall instruction	⚠ ⚠	Refit correctly	
8. Tyre rubbing against ▪ another part of the car ▪ its twin tyre	⚠ ⚠	Repair/Replace	

WARNING

Extremely dangerous. DO NOT drive your car in this condition.
You will be breaking the law and risking your life and the lives of others.

Very dangerous and may be illegal. Put right immediately.

TYRES - 5

Are your tyres in good condition?

WHAT THE MOT REQUIRES	HOW TO CHECK
Tread A tread pattern is the combination of plain surfaces and grooves extending across the breadth of the tread and round the entire circumference. The tread pattern excludes any tie-bars, tread wear indicators, or other features designed to wear out substantially before the remainder of the pattern, and other minor features. **Breadth of tread** The part of the tyre which can contact the road under normal conditions, measured at 90 degress to the line of the tread around the tyre. The tyre fitted to your car must have a 1.6mm tread depth, if your car was first used after 2 January 1933 and a. has not more than 8 passenger seats, exluding the driver's, or b. is a goods vehicle not exceeding 3500kg maximum gross weight	a. Raise wheels clear of ground b. Rotate each wheel slowly and check tread wear c. Check that tyre does not have re-cut* tread d. Check the tread pattern over the tyre. Check also that the tread depth meets the requirements using, as necessary, a depth gauge accepted for MOT testing **Central band** The area outside the central band can be completely bald and pass the test, but such wear is unlikely to happen

*'Re-cut'

- All or part of the original tread has been cut deeper or burned deeper, or
- A different tread has been cut or burned deeper than the original tread

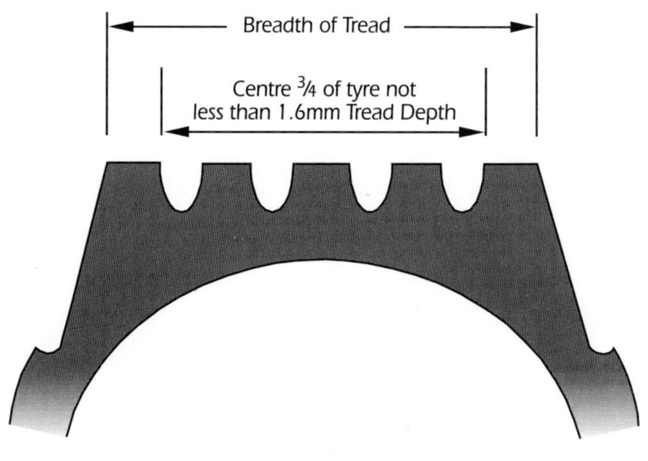

Are your tyres worn below the legal limit?

MAIN REASONS FOR FAILURE	REMARKS	ACTION	✓
1. The grooves of the tread pattern are not at least 1.6mm throughout a continuous band comprising ▪ the central three-quarters of the breadth of tread, and ▪ round all the tyre		Replace	
2. A recut tread (Illegal)		Replace	

WARNING

Extremely dangerous. DO NOT drive your car in this condition.
You will be breaking the law and risking your life and the lives of others.

Very dangerous and may be illegal. Put right immediately.

DRIVER'S VIEW OF THE ROAD - 1

Can you see the road ahead clearly?

WHAT THE MOT REQUIRES	HOW TO CHECK
The windscreen washers and wipers Your car must be fitted with one or more windscreen wipers and windscreen washers capable of cleaning the windscreen There is no requirement on the number of jets in the washers If your car has an opening windscreen there are no requirements for washers or wipers	1. Operate the windscreen washer/ wipers a. Check that the washers ▪ work correctly ▪ give out enough liquid to clean the windscreen with the help of the wipers b. Check that the wipers move over enough area of the windscreen to give the driver an adequate view of the road in front and forward of both sides of the car. 2. Check that the wipers are ▪ not loose ▪ in good condition

DRIVER'S VIEW OF THE ROAD - 2

Are your windscreen wipers/washers in good condition?

MAIN REASONS FOR FAILURE	REMARKS	ACTION	✓
1. a. A washer or wiper control ▪ missing, or ▪ not accessible to the driver	⚠⚠	Fit new control Replace control	
b. A wiper does not continue to work automatically when switched on	⚠⚠	Repair/Replace	
c. A wiper on the driver's side does not sweep over an area of windscreen large enough to give the driver a clear view ▪ through the windscreen ▪ of the road ahead ▪ forward of both sides of the car	⚠⚠	Adjust/Replace	
2. A wiper blade ▪ missing ▪ loose ▪ damaged or worn so that it cannot clear the windscreen enough to give the driver a clear view ▪ of the road ahead ▪ forward of both sides of the car	⚠⚠	Tighten/Replace	
3. The washers do not produce enough liquid to clear the windscreen with the help of the wipers.	⚠⚠	Clear jets/Top up reservoir	

WARNING

Very dangerous and may be illegal. Put right immediately.

DRIVER'S VIEW OF THE ROAD - 3

Can you see the road ahead clearly?

WHAT THE MOT REQUIRES	HOW TO CHECK
Windscreen The windscreen must give the driver a clear and uninterrupted view of the road ahead. **Stickers** Official stickers which are not readily removable – such as vehicle licences, parking permits and crime prevention stickers are only a reason for rejection if they seriously restrict the driver's view. **Repaired windscreens** An 'invisible' or barely detectable repair, finished flush with the surrounding glass, does not count as damage even if it exceeds the limit on damage allowed in the test. **Scratches** Scratches on the windscreen, ie light surface scratching, is not considered as damage. However, an area of concentrated scratching such as caused by the prolonged use of a defective wiper blade which obscures vision is likely to fail the test.	1. Examine the area swept by the wiper blades and the driver's view through it From the driver's seat, assess whether your view through the windscreen is reduced by ▪ damage to the windscreen ▪ anything attached to the swept area of the windscreen ▪ anything attached to the vehicle ▪ any modification to the vehicle

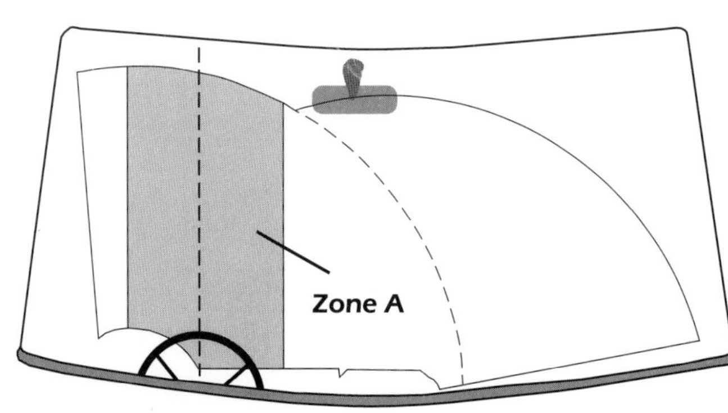

Fig 1. Zone A

Zone A

- 290mm wide
- centered on steering wheel
- Bounded at top by highest point of wipers sweep
- bounded at bottom by lowest point of wipers sweep

DRIVER'S VIEW OF THE ROAD - 4

Is your windscreen in good condition?

MAIN REASONS FOR FAILURE	REMARKS	ACTION	✓
1. In Zone A ▪ damage not contained within 10mm diameter circle, or ▪ a windscreen sticker or other obstruction encroaching more than 10mm ▪ a combination of minor damage areas which seriously restricts the driver's view	⚠⚠⚠	Repair/Replace/ Remove obstruction	
2. In the remainder of the swept area ▪ damage not contained within a 40mm diameter circle, or ▪ windscreen sticker or other obstruction encroaching more than 40mm	⚠⚠⚠	Repair/Replace/ Remove obstruction	

WARNING

Extremely dangerous. DO NOT drive your car in this condition.
You will be breaking the law and risking your life and the lives of others.

HORN - 1

Does your horn operate?

WHAT THE MOT REQUIRES	HOW TO CHECK
The horn Your car must be fitted with a HORN which can give audible and sufficient warning of your cars approach or position The sound must be continuous and uniform, and not strident ONLY SPECIFIED cars (eg. Police, Fire) may be fitted with a ▪ gong ▪ bell ▪ siren ▪ two-tone horn (except as an anti-theft device)	**Check your car is fitted with correct horn** a. Check that the horn can be easily sounded b. Operate the horn and make sure it gives out correct sound

HORN - 2

Is the correct horn fitted?

MAIN REASONS FOR FAILURE	REMARKS	ACTION	✓
1. No horn or horn control	⚠	Replace	
2. Horn control not accessible to driver	⚠	Repair	
3. Horn does not work	⚠	Repair/Replace	
4. Horn not loud enough	⚠	Repair/Replace	
5. Car fitted with a ▪ gong ▪ bell ▪ siren ▪ two-tone horn	⚠	Repair/Replace	
6. Tone not continuous or uniform	⚠	Repair/Replace	
7. Tone harsh or grating	⚠	Repair/Replace	

WARNING

Could also be dangerous and illegal.

EXHAUST - 1

Is the exhaust system in good condition?

WHAT THE MOT REQUIRES	HOW TO CHECK
The exhaust system Your car must be fitted with a SILENCER sufficient to reduce to a reasonable level the noise caused by the escape of the exhaust gases from the engine The EXHAUST SYSTEM must be free of major leaks.	1. Examine the entire exhaust system, including silencer(s) and mountings, check that exhaust is ▪ secure ▪ complete ▪ free of holes 2. With engine running, ▪ check for leaks ▪ assess noise emitted. Is it about right for the model of car? Repairs to the system are allowed providing the structure of the system is sound.

EXHAUST - 2

Is the correct exhaust fitted?

MAIN REASONS FOR FAILURE	REMARKS	ACTION	✓
1. Any part of exhaust system is missing or in poor condition	⚠ ⚠	Repair/Replace	
2. Any exhaust mounting ▪ missing ▪ not doing its job	⚠ ⚠	Repair/Replace	
3. A major leak	⚠ ⚠	Tighten/Replace	
4. A silencer which is excessively noisy due to its condition or type	⚠	Replace	

WARNING

Very dangerous and may be illegal. Put right immediately.

Could also be dangerous and illegal.

EXHAUST EMISSIONS (PETROL) - 1

Is your engine operating efficiently?

WHAT THE MOT REQUIRES	HOW TO CHECK
See also Appendix **This inspection applies to** All 4-stroke spark ignition (petrol) engined cars, light vans and pickups with four or more wheels, first used on or after 1 August 1975 But, note that ▪ kit cars ▪ amateur built vehicles, and ▪ Wankel rotary engined vehicles first used before August 1987 are to be considered as vehicles first used before August 1975. **Engine speed and temperature** During the check, the engine must be at its normal idle speed and normal operating temperature. Engine speed and temperature can be judged from experience or by referring to manufacturer's or other reliable data. **Gas analysis test** MOT testing stations must use a gas analyser for testing exhaust emissions. Since you are not likely to have access to such equipment, this information is for interest only.	1. a. raise the engine speed to around 2500 rpm or half maximum engine speed if this is lower. b. hold this speed steady for 20 seconds to ensure that the inlet and exhaust system is properly purged. c. allow the engine to return to idle and the emissions to stabilise. d. assess the smoke emitted from the tailpipe. **Gas analysis test** This test measures the amounts of carbon monoxide (CO) and hydrocarbons (HC) in the exhaust gas over a period of at least 5 seconds with the engine at idle. Hydrocarbons indicated when the analyser is sampling only clean air should be deducted from the HC reading obtained from the vehicle. If a vehicle meets the CO requirement at its normal idling speed but fails the HC check, the HC level must be rechecked with the engine at about 2000 rpm. If the HC reading is then 1200 ppm or less, the vehicle meets both the CO and HC requirements. The CO requirement must be met with the engine at its normal idling speed

EXHAUST EMISSIONS (PETROL) - 2

Are your exhaust emissions legal?

MAIN REASONS FOR FAILURE	REMARKS	ACTION	✓
1. The engine a. idles at a speed clearly above its normal idling speed b. emits dense blue or clearly visible black smoke for a continuous period of 5 seconds at idle **Note:** Older vehicles, particularly pre 1960, sometimes emit unavoidable smoke due to their design. This is **not** a reason for failure	⚠	Put right immediately	
2. a. cars first used on or after 1 August 1975 ▪ a HC content in the exhaust gas of more than 1200 ppm for a continuous period of 5 seconds ▪ a CO content in the exhaust gas of more than 6% for a continuous period of 5 seconds b. cars first used on or after 1 August 1983 ▪ a carbon monoxide content in the exhaust gas of more than 4.5% for a continuous period of 5 seconds	⚠	Put right immediately	

WARNING

Could also be dangerous and illegal.

EXHAUST EMISSIONS (DIESEL) - 3

Is your engine operating efficiently?

WHAT THE MOT REQUIRES	HOW TO CHECK
This inspection applies to All compression ignition (diesel) engined vehicles **Engine speed and temperature** During the check, the engine must idle at its normal idle speed and be at its normal operating temperature. Engine speed and temperature can be judged from experience or by referring to manufacturer's or other reliable data. **Smoke emission test** MOT testing stations must use a smoke meter for testing exhaust emissions. Since you are not likely to have access to such equipment, this information is for interest only. The smoke meter is used to measure the density of the exhaust emission averaged over several applications of the accelerator.	1. **Vehicles first used before 1 August 1979** a. raise the engine speed to around 2500rpm b. hold this speed steady for 20 seconds to ensure that the system is properly purged c. allow the engine to return to idle and the emissions to stabilize d. observe the smoke emitted 2. **Vehicles first used on or after 1 August 1979** a. raise the engine speed to around 2500 rpm b. hold this speed steady for 20 seconds to ensure that the system is properly purged c. raise the engine speed slowly to maximum to check the operation of the governor d. allow the engine to return to idle e. depress the accelerator sharply to maximum then release immediately f. repeat this operation and observe the smoke emitted **Note:** If the smoke emitted is a black haze or darker, the vehicle might not pass the test – if in doubt seek expert advice.

EXHAUST EMISSIONS (DIESEL) - 4

Is your exhaust emission within the requirements?

MAIN REASONS FOR FAILURE	REMARKS	ACTION	✓
1. For all vehicles **first used before 1 August 1979** the engine emits dense blue or clearly visible black smoke for a period of 5 seconds at idle		Put right immediately	
2. After the required free accelerations have been carried out, the mean of the last three smoke levels is ▪ For a non-turbocharged engine of a vehicle **first used on or after 1 August 1979**, more than 3.2m-1 ▪ For a turbocharged engine of a vehicle **first used on or after 1 August 1979**, more than 3.7m-1 ▪ The exhaust emits excessive smoke or vapour of any colour to an extent likely to obscure vision		Put right immediately	
3. Exhaust emits excessive smoke or vapour of any colour to an extent likely to obscure the vision of other road users.		Put right immediately	

WARNING

Could also be dangerous and illegal.

STRUCTURE - 1

Is the structure of your car safe?

WHAT THE MOT REQUIRES	HOW TO CHECK
The structure The STRUCTURE of your car must be sound and secure and must not cause danger to any person in the car or on the road The effect of corrosion on the safety of a vehicle depends on ▪ the extent of the corrosion ▪ the function of the section on which the corrosion occurs. A little corrosion on an important load-bearing part of a car can make a vehicle unsafe when it destroys the continuity of the load bearing structure. On the other hand, heavy corrosion of unimportant sections might have no effect on the car's safety. Corrosion of a particular part, such as a body sill, might be very important on one type of car, but less important on another. In figures 1 to 4, the shaded parts show the important load bearing parts of typical cars.	1. Preferably with car over a pit or on a raised hoist, check that there is no ▪ fracture ▪ damage ▪ corrosion on car structure that could adversely affect the correct functioning of a. braking system b. steering gear For specific areas to check, see the diagrams below and on the following pages

Figure 1

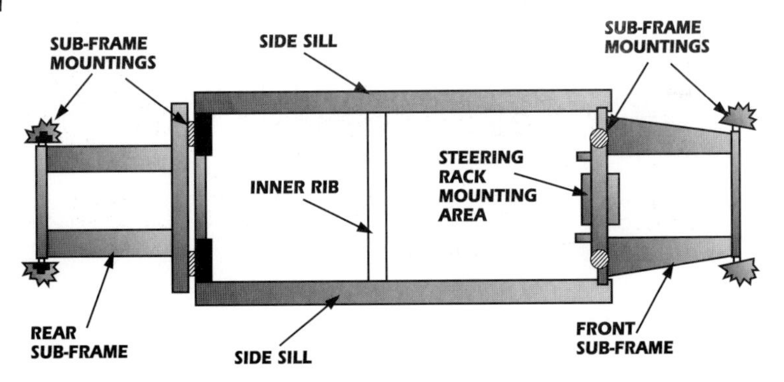

Is the structure in good condition?

MAIN REASONS FOR FAILURE	REMARKS	ACTION	✓
1. Any part ▪ cracked ▪ damaged ▪ corroded and likely to affect the working of the steering or brakes.	⚠⚠⚠	Repair/Replace	

WARNING

Extremely dangerous. DO NOT drive your car in this condition.
You will be breaking the law and risking your life and the lives of others.

Figure 2

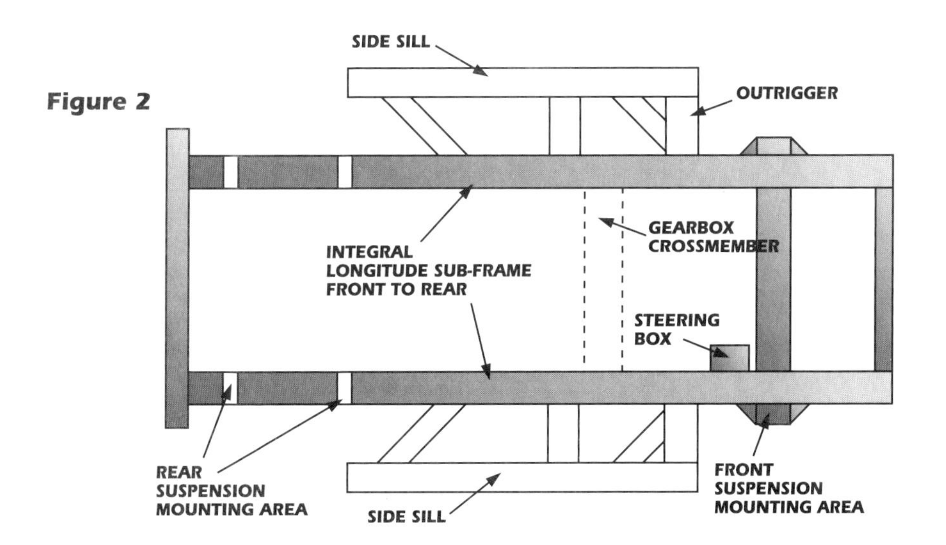

STRUCTURE - 3

Is the structure of your car safe?

WHAT THE MOT REQUIRES	HOW TO CHECK
The STRUCTURE of your car must be sound and secure and must not cause danger to any person in the car or on the road. **Prescribed areas** The tester will look in particular for corrosion in prescribed areas. These are ▪ load bearing points to which testable items are attached ▪ areas of supporting structure within 30cm of mounting points	1. Identify ▪ important load-bearing areas ▪ mounting points 2. Check for corrosion. Press hard and note any 'give' or disintegration. DO NOT ▪ use a sharp instrument ▪ 'dig' at structure ▪ use heavy impact blows 'Bodged'. (especially non-metal) repairs can be found by ▪ using a magnet, or ▪ noting dull sound when tapped

Figure 3
Topside of typical car body of monocoque construction

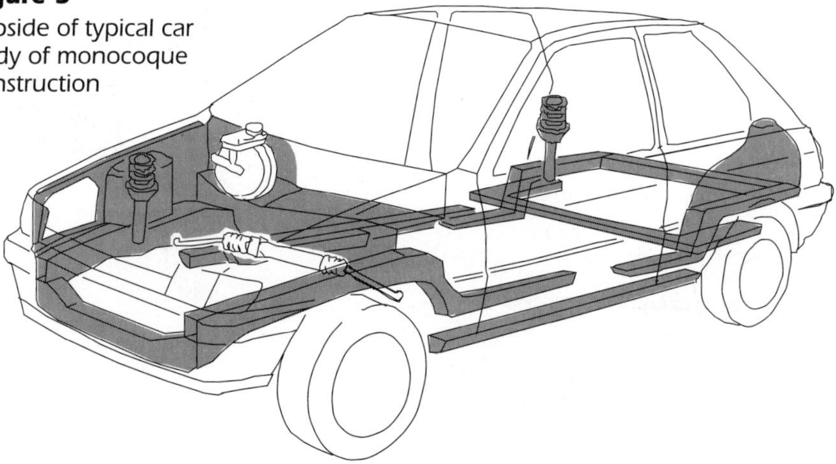

STRUCTURE - 4

Is the structure in good condition?

MAIN REASONS FOR FAILURE	REMARKS	ACTION	✓
1. Excessive ▪ corrosion ▪ cracks ▪ damage on a load bearing area within 30cm of a mounting point		Repair	

WARNING

Extremely dangerous. DO NOT drive your car in this condition.
You will be breaking the law and risking your life and the lives of others.

Figure 4

Underside of typical car body of monocoque construction. (Doors and front wings omitted)

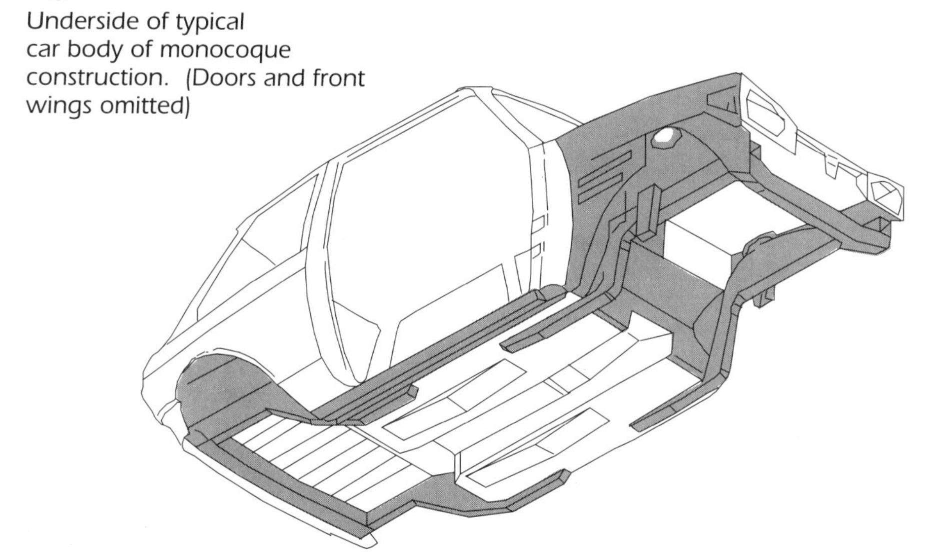

Reproduced by kind permission of the Motor Industry Repair Research Centre (M.I.R.R.C)

BODY CONDITION - 1

Is the body in good condition?

WHAT THE MOT REQUIRES	HOW TO CHECK
The body of your car must not be in such condition as to cause danger to any person in the car or to other road users.	**Body Condition** 1. Check the body for sharp edges or projections caused by corrosion or damage. **Driver or passenger security** 2. Open and close ▪ the driver's door ▪ all passenger doors 3. Check that all doors can be ▪ opened from both inside and outside the car ▪ latched securely in the closed position **Load security** 4. Check that each of the following ▪ is secured in the closed position, or ▪ can be secured in the closed position a. boot lid b. tailgate c. loading door d. hinged tailboard e. dropside **Spare wheel and carrier** 5. Check that any externally fitted spare wheel and/or carrier is securely attached to the vehicle.

BODY CONDITION - 2

Is the body of your car likely to endanger others?

MAIN REASONS FOR FAILURE	REMARKS	ACTION	✓
1. A sharp edge which makes the car dangerous to other road users	⚠⚠⚠	Repair	
2. A driver's or front passengers door cannot be opened from both inside and outside the vehicle, or any door that does not latch securely in the closed position	⚠⚠⚠	Repair	
3. One of the following ▪ not secured in the closed position, or ▪ cannot be secured in the closed position a. boot lid b. tailgate c. loading door d. hinged tailboard e. dropside	⚠⚠⚠	Repair	
4. An externally fitted spare wheel or carrier insecure to the extent that it is likely to fall off	⚠⚠⚠	Repair	

WARNING

Extremely dangerous. DO NOT drive your car in this condition.
You will be breaking the law and risking your life and the lives of others.

BODY CONDITION - 3

Is the body secure?

WHAT THE MOT REQUIRES	HOW TO CHECK
Vehicles with separate bodies have extra requirements. The body ▪ must be securely attached to its supporting members and to the chassis, and ▪ must not be likely to cause a danger to other road users **Seat security** The driver's and front passenger must be securely fitted to the vehicle. The back-rests of **all** seats must be capable of being secured in their normal upright position.	1. Check a. all fixings (e.g. brackets) securing the body to the chassis b. all fixings (e.g. brackets) securing the body to a sub-frame c. all securing bolts, rivets or welds on the fixings for ▪ presence ▪ security ▪ fracture or distortion ▪ excessive wear ▪ damage **Note:** The tester will assess the cumulative affect of loose, defective or missing bolts, rivets or welds. **Seat security** Check that ▪ The driver's seat ▪ The front passengers seat are fully secure to the vehicle Check that the back-rests of ALL SEATS can be secured in their normal upright position.

BODY CONDITION - 4

Are the front seats secure?

MAIN REASONS FOR FAILURE	REMARKS	ACTION	✓
1. The body has shifted in relation to the chassis and this is likely to cause loss of control of the vehicle.	⚠⚠⚠	Secure	
2. The body is not securely attached to its supporting members and this could endanger other road users.	⚠⚠⚠	Secure	
3. The driver's seat or the front passengers seat is not secure.	⚠⚠⚠	Secure	
4. The back-rest of any seat cannot be secured in its normal upright position.	⚠⚠⚠	Secure	

WARNING

Extremely dangerous. DO NOT drive your car in this condition.
You will be breaking the law and risking your life and the lives of others.

SEAT BELTS - 1

Are the correct seat belts fitted?

WHAT THE MOT REQUIRES	HOW TO CHECK
Your car must be fitted with seat belts on the driver's and the passenger's seats. For details on the seat belts required, see pages 96 – 101. The tester will examine the required seat belts only. Additional seat belts will not be tested. However, on most cars all fitted seat belts are required.	1. Check that each seat required to have a seat belt fitted has one of the correct type (see pages 96 – 101) 2. Pull each seat belt away from the anchorages to check that the anchorage is secured to the car body You might not be able to do this if your car has seat belts where the fixing secures to the seat (integral seat belts) 3. Check webbing for cuts and signs of deterioration, especially around anchorages, buckles and loops 4. Fasten each seat belt lock. Try to pull locked sections apart Check that the lock releases when required 5. Check the attachment fitting and adjuster for ▪ fractures, or ▪ deterioration

Are the correct seat belts in good condition?

MAIN REASONS FOR FAILURE	REMARKS	ACTION	✓
1. A seat belt ▪ not fitted ▪ of the wrong type		Replace	
2. A seat belt not securely fitted ▪ to the structure of the car ▪ to the seat (a bolt loose) For integral seat belts, the seat is not securely attached to the floor		Repair/Replace	
3. On the webbing a. A cut which causes the fibres to separate b. Fluffing or fraying ▪ enough to weaken the belt ▪ interferes with the correct working of the belt c. Stitching ▪ frayed ▪ incomplete ▪ repaired		Tighten/Replace	
4. Locking mechanism does not lock the belt when the belt is pulled suddenly		Replace	
5. An attachment or adjuster fractured or badly deteriorated		Replace	

WARNING

Very dangerous and may be illegal. Put right immediately.

SEAT BELTS - 3

Are the seat belts safe?

WHAT THE MOT REQUIRES	HOW TO CHECK
The car structure within 30cm of ▪ a seat belt anchorage point, or ▪ if the seat has an integrated belt, a seat mounting must be free of corrosion and fractures	6. Check flexible buckle stalks for ▪ corrosion ▪ deterioration (it must have no broken strands) 'Waggle' stalks and listen for clicking noise indicating broken strands or cable 7. With retracting seat belts, pull a section of webbing from the retractor and then let go. Some types might need help before they retract You might not be able to do this if your car has integral seat belts 8. Check the vehicle structure around seat belt mounting points for corrosion. You might have to examine floor-mounted anchorage points from underneath the car 9. If the seats have integral seat belts, check the structure around the seat mountings

SEAT BELTS - 4

Are the seat belts securely attached?

MAIN REASONS FOR FAILURE	REMARKS	ACTION	✓
6. A flexible stalk ▪ corroded ▪ deteriorated so that it is likely to fail	⚠⚠⚠	Replace	
7. A seat belt retractor does not retract the belt enough to restrain the wearer	⚠⚠⚠	Repair/Replace	
8. On the car's structure ▪ excessive corrosion ▪ distortion ▪ fracture on a load bearing area within 30cm of ▪ a seat belt mounting ▪ a seat mounting, when the seat has integral belts	⚠⚠⚠	Repair	

WARNING

Extremely dangerous. DO NOT drive your car in this condition.
You will be breaking the law and risking your life and the lives of others.

SEAT BELTS - 5

Seat belt requirements for vehicles **first used before 1 April 1987**

VEHICLE DESCRIPTION

1. **Passenger vehicles**
   - with 4 or more wheels
   - with upto 12 passenger seats
   - first used on or after 1 January 1965
2. **3-wheeled vehicles**
   - with an unladen weight over 410Kg first used on or after 1 January 1965, or
   - with an unladen weight over 255Kg if first used on or after 1 September 1970

   Except vehicles
   - less than 410Kg unladen equipped with a driver's seat of a type requiring the driver to sit astride it, and
   - constructed or assembled by a person not ordinarily engaged in the trade or business of manufacturing vehicles of this type
3. **Goods vehicles, motor caravans and ambulances**
   - with an unladen weight **not exceeding 1525Kg**
   - first used on or after 1 April 1967
4. **Goods vehicles, motor caravans and ambulances**
   - with a design gross weight **not exceeding 3500Kg**
   - first used on or after 1 April 1980

   except those first used before 1 April 1982, if they are of a model manufactured before 1 October 1979 with an unladen weight exceeding 1525Kg.

SEAT POSITION

Driver's and "Specified Front Passenger's" Seat (See Note 1, page 98)	Centre Front Seat	Forward Facing Rear Seats
A. Vehicles first used before 1 April 1981: A belt which restrains the upper part of the body (but need not include a lap belt) for each seat. B. Vehicles first used after 31 March 1981: A 3 point (lap/diagonal) belt (see Note 2, page 98)	No requirement	No requirement

Seat belt requirements for vehicles **first used after 31 March 1987**

VEHICLE DESCRIPTION

1. Passenger vehicles and dual purpose vehicles with not more than 8 passenger seats, except 3 wheeled vehicles
   - with an unladen weight of 255Kg or less
   - with an unladen weight over 255Kg but less than 410Kg
     - equipped with a driving seat of a type requiring the driver to sit astride it, and
     - constructed or assembled by a person not ordinarily engaged in the trade or business of manufacturing vehicles of this type

SEAT POSITION

Driver's and "Specified Front Passenger's" Seat (See Note 1, page 98)	Centre Front Seat	Forward Facing Rear Seats
3 point belts for each seat. (See Note 2, page 98)	3 point belt, lap belt or a disabled person's belt.	A. Vehicles with not more than 2 rear seats: Either i. A 3 point inertia reel belt for at least one seat; or ii. A 3 point belt, lap belt, disabled person's belt or child restraint for each seat. B. Vehicles with more than 2 rear seats: Either i. A 3 point inertia reel belt on an outboard seat and a 3 point static or inertia reel belt, lap belt, disabled persons belt or child restraint for at least one other seat; or ii. A static 3 point belt for one seat and a disabled person's belt or child restraint for at least one other seat; or iii. A 3 point belt, lap belt, disabled person's belt or child restraint for each seat. See additional information on pages 99 - 101.

Seat belt requirements for vehicles **first used after 31 March 1987**

VEHICLE DESCRIPTION	SEAT POSITION: Driver's and "Specified Front Passenger's" Seat (See Note 1 below)	SEAT POSITION: Centre Front Seat	SEAT POSITION: Forward Facing Rear Seats
2. Goods Vehicles	3 point belts for each seat. (See Note 2, below)	3 point belt, lap belt or a disabled person's belt.	No requirement
3. Vehicles first used before 1 October 1988 which are: ▪ minibuses with up to 12 passenger seats ▪ motor caravans and ambulances with a design gross weight not exceeding 3500kg	As above	No requirement	No requirement
4. Minibuses, motor caravans and ambulances ▪ with a design gross weight not exceeding 3500Kg ▪ first used after 30 September 1988	As above	3 point belt or a lap belt	No requirement

Note 1: The "specified front passenger seat" requiring a seat belt is the seat which is

- Foremost in the vehicle, and
- Furthest from the driver's seat

unless there is a fixed partition separating the passenger seat from a space in front of it which is alongside the driver's seat, eg certain types of taxis, buses etc.

Note 2: '3 point belt' means a seat belt which

i. restrains the upper and lower parts of the torso
ii. includes a lap belt
iii. is anchored at not less than three points, and
iv. is designed for use by an adult

One or two forward facing rear seats

Vehicles first used on or after 31 March 1987. Forward facing rear seats must have **at least** the type and number of seat belts shown below.

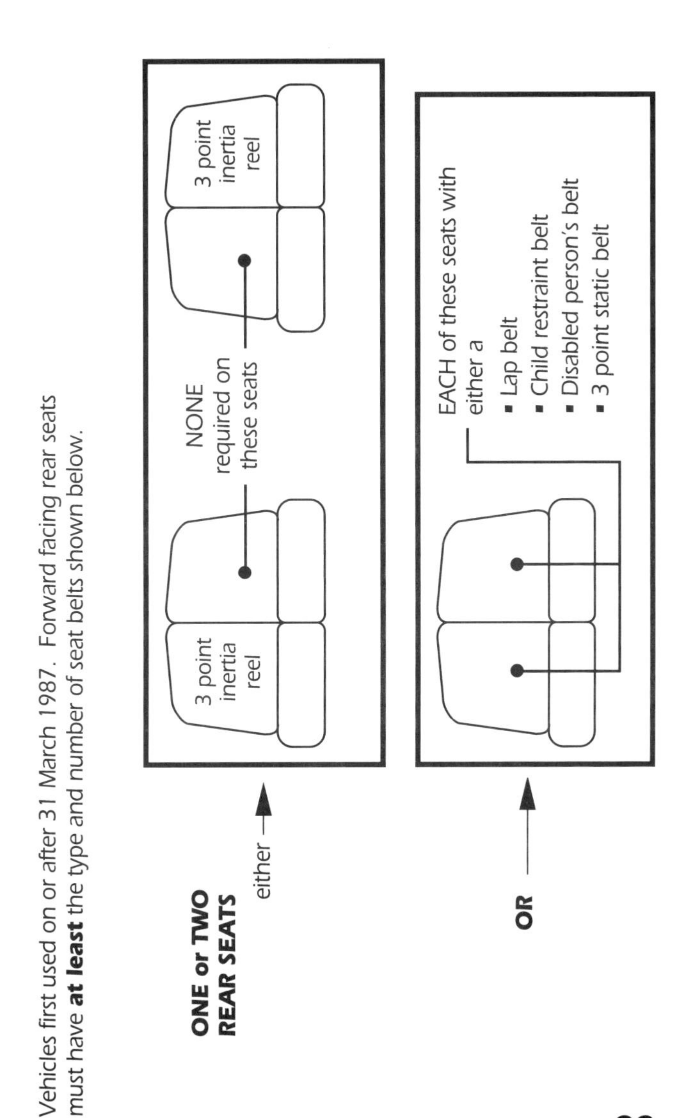

SEAT BELTS - 9

Three forward facing rear seats

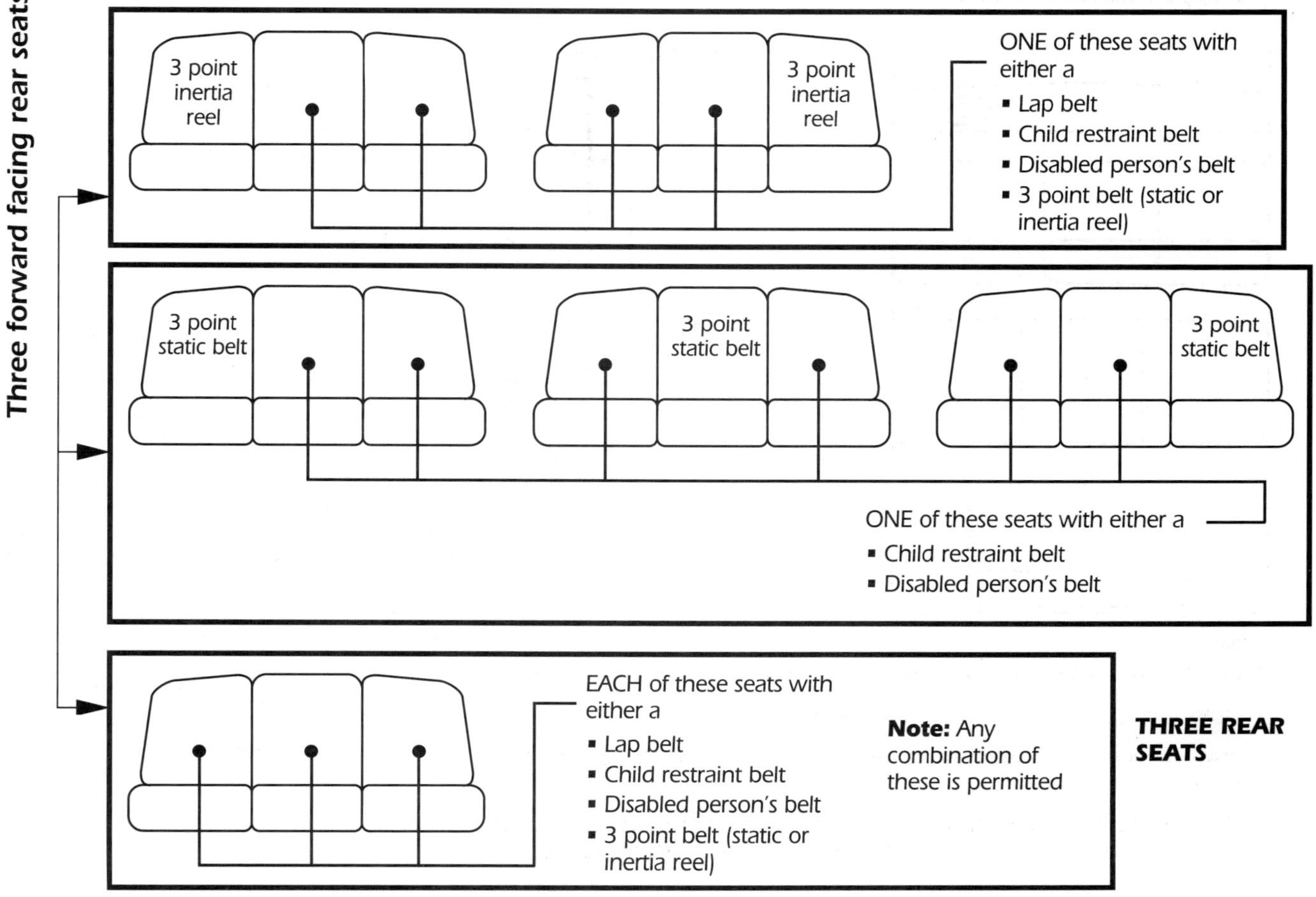

More than three forward facing rear seats

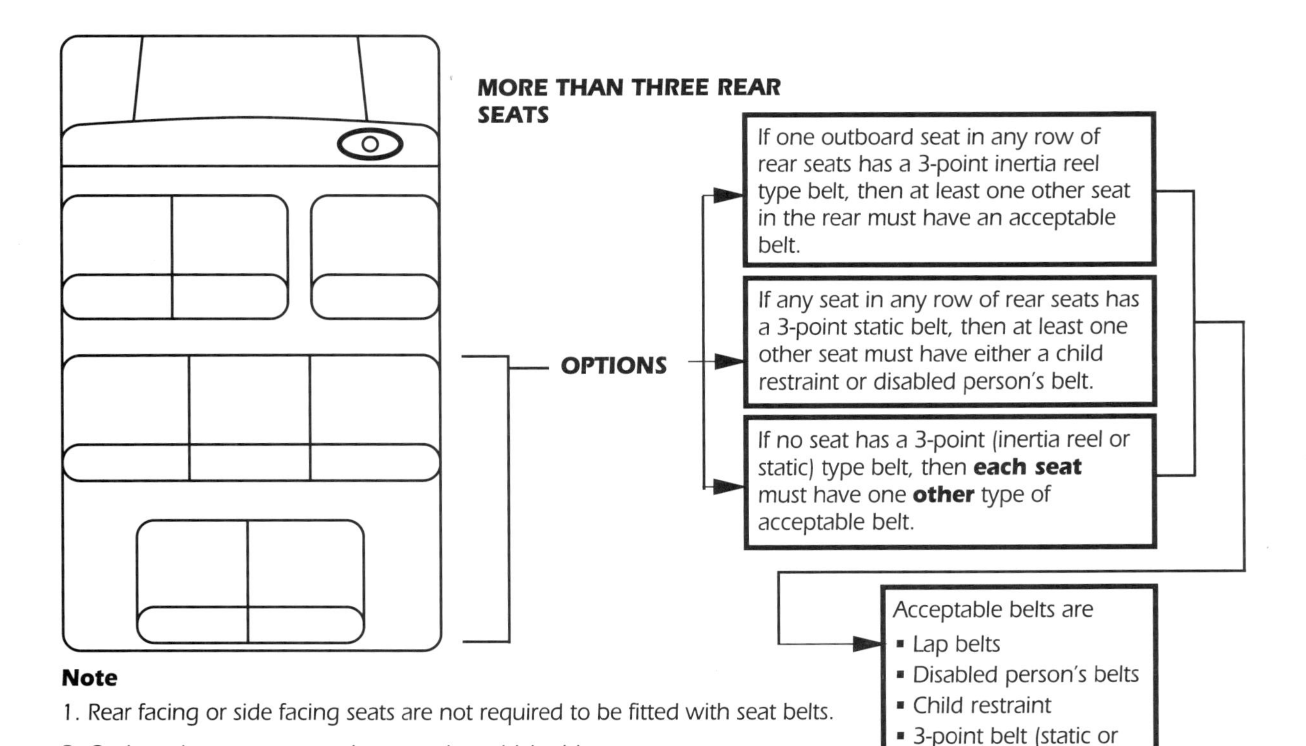

Note

1. Rear facing or side facing seats are not required to be fitted with seat belts.
2. Outboard seats are seats closest to the vehicle sides.
3. Rear seat belts are not required for vehicles with more than 8 passenger seats in any configuration.
4. Occasional seats fitted in the rear of extendable limousines which fold when not in use are not required to be fitted with seat belts or anchorage points.

NUMBER PLATES & VIN - 1

Is your car registration number displayed correctly?

WHAT THE MOT REQUIRES	HOW TO CHECK
Number plates Your car must display number plates ▪ one visible from the front of the car ▪ one visible from the rear of the car. The registration plate must be set out in one of two ways (see below) **Layout 1** Letters on one line, figures on another **Layout 2** The gap between groups of letters and figures (Gap A) must be at least double the gap between individual letters and figures (Gap B) **Vehicle Identification Number (VIN or chassis number)** If your car was first used on or after 1 August 1980, it must have a vehicle identification number fitted.	1. Check that both number plates are ▪ present ▪ in good condition ▪ secure 2. Check that both number plates are fully legible, one on the front of the car and one on the rear. 3. Check that the registration number (letter and figures) are correctly ▪ formed ▪ spaced Numbers must not be capable of being misread because of ▪ incorrect shape ▪ incorrect position ▪ badly positioned or uncovered retaining bolts 4. Check that your car is fitted with a legible vehicle identification number

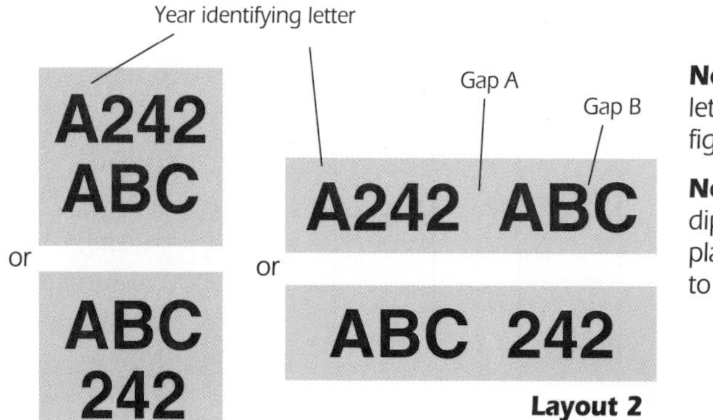

Note 1: A year identifying letter is regarded as a figure

Note 2: Foreign or diplomatic registration plates need not conform to these layouts

Fig 1. Permitted numberplate layouts

NUMBER PLATES & VIN - 2

Is your car's vehicle identification number displayed?

MAIN REASONS FOR FAILURE	REMARKS	ACTION	✓
1. A number plate ▪ missing ▪ letter or figure missing or incomplete ▪ so insecure that it is likely to fall off	⚠⚠	Replace Tighten/Replace retaining bolts	
2. A number plate faded, dirty, deteriorated or obscured (eg. by a towbar), so that it is likely to be misread or is not easily legible by a person standing approximately 20m to the front/rear of the vehicle	⚠⚠	Replace	
3. A number plate displays ▪ letters/figures incorrectly spaced ▪ likely to be misread	⚠⚠	Remove obstruction, or reposition number plate	
4. A vehicle identification number is ▪ not fitted ▪ not legible	⚠⚠	Seek advice from Vehicle Registration Office	

WARNING

Very dangerous and may be illegal. Put right immediately.

MIRRORS - 1

Does your car have the correct mirrors fitted?

WHAT THE MOT REQUIRES	HOW TO CHECK
Obligatory mirrors Obligatory mirrors are a. An exterior mirror fitted to the offside (right-hand side when seated in the driver's seat), or b. An exterior mirror fitted to the nearside (left-hand side when seated in the driver's seat), or c. An interior mirror **Vehicles requiring only one mirror** Passenger vehicles with no more than 7 passenger seats first used before 1 August 1978 must have any one of the above options. **Vehicles requiring two mirrors** The following vehicles must have TWO mirrors, one of which must be option 'a' ▪ passenger vehicles of any age with more than 7 seats ▪ all passenger vehicles first used on or after 1 August 1978 (NOT a minibus) ▪ all goods vehicles **Minibuses (vehicles constructed or adopted to carry more than 8 but not more than 16 seated passengers)** Minibuses first used on or after 1 April 1983 must have an external mirror fitted on both the offside and nearside ('a' and 'b') **Additional mirrors** Additional mirrors are not part of the test.	1. Check that your car has the correct combination of obligatory mirrors 2. Check that the obligatory mirrors are ▪ secure ▪ in good condition ▪ are clearly visible from the driver's seat, or can be adjusted to be clearly visible ▪ give a clear view to the rear of the vehicles from the driver's seat.

Are your mirrors doing their job?

MAIN REASONS FOR FAILURE	REMARKS	ACTION	✓
1. An obligatory mirror missing		Fit correct mirrors	
2. An obligatory mirror			
▪ loose		Tighten mirror	
▪ damaged or deteriorated so that the view to the rear is impaired		Replace	
▪ not clearly visible, or cannot be adjusted to be clearly visible, from the driver's seat		Replace	
▪ does not give a clear view to the rear from the driver's seat		Replace	

WARNING

Extremely dangerous. DO NOT drive your car in this condition.
You will be breaking the law and risking your life and the lives of others.

Very dangerous and may be illegal. Put right immediately.

FUEL SYSTEM - 1

Is your fuel system in good condition?

WHAT THE MOT REQUIRES	HOW TO CHECK
The fuel system in your car must be secure, free of leaks and not cause any danger to other road users.	Check for insecurity and leaks a. fuel tank b. all visible hoses, pipes, unions c. all visible fuel system components Check the ▪ fuel cap for fit ▪ condition of the sealing washer and flange

Is your fuel system safe?

MAIN REASONS FOR FAILURE	REMARKS	ACTION	✓
1. Fuel leaking	⚠⚠⚠	Repair immediately	
2. An insecure fuel system component	⚠⚠⚠	Repair immediately	
3. Fuel cap does not fasten securely	⚠⚠⚠	Replace fuel cap	
4. Fuel cap sealing washer or mounting flange damaged or deteriorated so that fuel is likely to escape, for example when vehicle is cornering	⚠⚠⚠	Replace fuel cap	

WARNING

Extremely dangerous. DO NOT drive your car in this condition.
You will be breaking the law and risking your life and the lives of others.

APPENDIX - EXHAUST EMISSIONS PETROL

The Exhaust Emission Test

A test of vehicle exhaust emissions is part of the MOT test for all 4-stroke spark ignition engined vehicles with 4 or more wheels in Classes IV and VII: cars, light vans and pickups.

Two exhaust gases are included

- Carbon Monoxide (CO)
- Hydrocarbons (HC)

Assessment on most vehicles is straightforward, but some factors should be borne in mind.

Conducting the Test

- The test must be conducted with the engine warm. Testing a cold engine could lead to an unjustified failure.
- Any enrichment device must be switched off.
- The engine must be idling normally during the test and must not have any significant electrical loading such as heated seats or heated rear windows.

If an engine will not idle, an assistant may apply light throttle pedal pressure.

To assess that these conditions are met, MOT testers can either

- Use their own judgement, or
- Refer to manufacturer's or other reliable data

Electric engine-cooling fans

Many modern vehicles have electric engine-cooling fans which can cut in during an emission test. The extra load on the alternator reduces the idle speed and causes the engine management system to react. This gives rise to highly variable readings.

If this happens during a test, the tester must not continue until the fan switches off and the readings settle down.

Unstable readings

Some vehicles give unstable readings due, for example, to their carburettor or fuel injection system design. Before failing a vehicle, the tester must establish that a particular limit has been exceeded for a continuous period of 5 seconds.

Holed exhaust

A holed exhaust can allow air to be sucked in, causing artificially low readings.

Where a vehicle has an exhaust holed to the extent that MOT failure is justified, the emissions should be rechecked when the exhaust is repaired even if the vehicle does not leave the testing station in the meantime. The driver should be told that any emission readings taken with a leaking exhaust might be incorrect.

Holes not justifying MOT failure do not normally have a significant effect on the composition of the exhaust gases at the tailpipe and can be ignored.

Total gas emitted

The MOT limits laid down relate to the total exhaust gas being emitted by the vehicle.

If a vehicle has a dual exhaust system, the emissions from the tailpipes must be averaged. This is done by adding together the readings and dividing by two, eg

1st pipe emits 6% CO, 400 ppm HC
2nd pipe emits 4% CO, 500 ppm HC

Average CO reading is

$$\frac{6 + 4}{2} = 5\%$$

Average HC reading is:

$$\frac{400 + 500}{2} = 450 \text{ ppm}$$

Single exhaust systems

A single exhaust system has at least one point in the system where all the exhaust gases from the engine travel through the same pipe, even though the system might split at some point to separate silencers or tailpipes. Only one of these need be checked.

Dual exhaust system

A dual exhaust system has two separate pipes from the engine manifold all the way back to the tailpipes. An exhaust system with a balance tube between separate pipes is still considered a dual exhaust.

Multi fuel vehicles

Vehicles which run on more than one fuel (eg petrol and LPG) should be tested on the fuel they are running on when presented.

There is a slight difficulty with LPG vehicles: the hydrocarbons emitted are propane rather than hexane. So the HC reading obtained must be divided by the "propane/hexane equivalency factor" (PEF) marked on the gas analyser. For example:

An LPG vehicle gives a HC reading of 700 ppm.
The PEF marked on the machine is 0.48.

So the actual MOT value is

$$\frac{700}{0.48} = 1458 \text{ (fail)}$$

Some exhaust gas analysers do this automatically.

Vehicles which only just pass

Cars equipped with catalytic converters are subject to the exhaust emission test, although they normally run at levels well below the MOT limits. Many modern vehicles also run normally well below the MOT limits.

Where such a vehicle barely passes the MOT test, but the tester knows that it is capable of more efficient operation, the owner should be told. Vehicles should be tuned to the manufacturer's recommended settings wherever possible, but tuning is **not** part of the MOT test.

Vehicles which are incapable of passing

Regulations do not require vehicles to achieve CO or HC readings below the original capability of the engine when new. A very few vehicles might never have been able to meet the MOT limits.

Where a vehicle owner claims that this is the case, and has sound supporting evidence (eg a letter from the vehicle manufacturer), the vehicle should be exempted from the CO and HC emission test.

If the owner **does not have** sound supporting evidence, a Test Certificate **should be refused**.

IF YOUR CAR FAILS

If your car has failed the test, please read these notes

1. Your car does not meet the legal requirements. If you intend to continue to use it on the road, you should have it repaired WITHOUT DELAY.
2. You will be committing an offence if you use the car on the road if it does not have a current test certificate, except when
   - it is not of a testable age, or when you are
     - taking it to a testing station for a test BOOKED IN ADVANCE
     - bringing it away from a testing station after it has failed the test
     - taking it to (or from) a place where by PREVIOUS ARRANGEMENT repairs are to be done to remedy the defects for which the car was failed.

Even in these circumstances, you can still be prosecuted if your car is not roadworthy under the various regulations affecting its construction and use. Also, the insurance may not cover you to drive the car.

HOW TO APPEAL

If your car fails its MOT Test, you may if you wish appeal to the Vehicle Inspectorate's local Enforcement Area Office using form VT17. MOT Testing stations must give you the address of the office and copies of the form.

You must pay a fee which you get back if your appeal succeeds. The completed form (the notice of appeal) and the fee must be received by the Vehicle Inspectorate's Enforcement Area Office within 14 days of the test.

DO NOT repair or alter the items which are the subject of your appeal before the Inspectorate has examined them. If you do repair or alter them the outcome of the appeal may be affected.

Printed in the United Kingdom for HMSO
Dd 296176 C60 3/94 531/3 12521